Making Language in the University

NEW PERSPECTIVES ON LANGUAGE AND EDUCATION
Founding Editor: **Viv Edwards,** *University of Reading, UK*
Series Editors: **Phan Le Ha,** *University of Hawaii at Manoa, USA* and **Joel Windle,** *Monash University, Australia.*

Two decades of research and development in language and literacy education have yielded a broad, multidisciplinary focus. Yet education systems face constant economic and technological change, with attendant issues of identity and power, community and culture. This series will feature critical and interpretive, disciplinary and multidisciplinary perspectives on teaching and learning, language and literacy in new times.

All books in this series are externally peer-reviewed.

Full details of all the books in this series and of all our other publications can be found on http://www.multilingual-matters.com, or by writing to Multilingual Matters, St Nicholas House, 31-34 High Street, Bristol BS1 2AW, UK.

NEW PERSPECTIVES ON LANGUAGE AND EDUCATION: 82

Making Language Visible in the University

English for Academic Purposes and Internationalisation

Bee Bond

MULTILINGUAL MATTERS
Bristol • Blue Ridge Summit

DOI https://doi.org/10.21832/BOND9295
Library of Congress Cataloging in Publication Data
A catalog record for this book is available from the Library of Congress.
Names: Bond, Bee – author.
Title: Making Language Visible in the University: English for Academic Purposes and Internationalisation/Bee Bond.
Description: Bristol, UK; Blue Ridge Summit: Multilingual Matters, 2020. | Series: New perspectives on Language and Education: 82 | Includes bibliographical references and index. | Summary: 'This book focuses on the nexus of language, disciplinary content and knowledge communication against the background of Higher Education's current push for internationalisation. It has an emphasis throughout on the practice of teaching and the barriers and enablers to that practice within a particular context' – Provided by publisher.
Identifiers: LCCN 2020016675 (print) | LCCN 2020016676 (ebook) | ISBN 9781788929288 (paperback) | ISBN 9781788929295 (hardback) | ISBN 9781788929301 (pdf) | ISBN 9781788929318 (epub) | ISBN 9781788929325 (kindle edition)
Subjects: LCSH: English language – Study and teaching (Higher) – Foreign speakers. | Academic writing – Study and teaching (Higher)
Classification: LCC PE1128.A2 B58 2020 (print) | LCC PE1128.A2 (ebook) | DDC 428.0071/1 – dc23
LC record available at https://lccn.loc.gov/2020016675
LC ebook record available at https://lccn.loc.gov/2020016676

British Library Cataloguing in Publication Data
A catalogue entry for this book is available from the British Library.

ISBN-13: 978-1-78892-929-5 (hbk)
ISBN-13: 978-1-78892-928-8 (pbk)

Multilingual Matters
UK: St Nicholas House, 31-34 High Street, Bristol BS1 2AW, UK.
USA: NBN, Blue Ridge Summit, PA, USA.

Website: www.multilingual-matters.com
Twitter: Multi_Ling_Mat
Facebook: https://www.facebook.com/multilingualmatters
Blog: www.channelviewpublications.wordpress.com

The policy of Multilingual Matters/Channel View Publications is to use papers that are natural, renewable and recyclable products, made from wood grown in sustainable forests. In the manufacturing process of our books, and to further support our policy, preference is given to printers that have FSC and PEFC Chain of Custody certification. The FSC and/or PEFC logos will appear on those books where full certification has been granted to the printer concerned.

Typeset by Riverside Publishing Solutions.

Contents

Acknowledgements vii
Preface ix

Introduction: Contextualising the Problem, Defining Terms 1

Aims and Purposes 1
What is Inclusive Education? 2
Internationalisation as a Driver for Change 3
Who is an International Student? 4
The Position of English for Academic Purposes 7

1 The Accidental Scholar 11

The Scholarship of Learning and Teaching 12
The Project 18
Methodological Approach 20
Project Design and Data Collection 22
Data Analysis 25
Contexts 27
Participants 29
Chapter Summary and Practical Lessons Learned for SoTL 33

2 Tracing a Student Journey: The Stories of Mai and Lin 35

Mai 36
Lin 45
Chapter Summary 51

3 The Taught Post-Graduate Curriculum 52

What is a Curriculum? 53
What is the Purpose of a Taught Post-Graduate Programme? Who is it For? 56
Changing Identities 62
Temporality 71
The Importance of Trust and Emotion 75

Making Agential Decisions 83
Chapter Summary and Practical Lessons Learned: TPG Education 87

4 **Language and the Academic Curriculum** 90
Language and Content Knowledge 92
Language and Assessment Practices 106
Language and Pedagogy 120
Chapter Summary and Practical Lessons Learned: Focusing on Language in Content Teaching 126

5 **Language and Academic Norms** 128
Language and Cultural Capital 128
Language and Social Capital 137
Language as a Threshold Concept 142
Language as Tacit Knowledge 144
Chapter Summary and Practical Lessons Learned: A Language Connected Curriculum 147

6 **The Place of English for Academic Purposes** 150
Teacher or Academic? 151
(Denial of) Agency 157
Developing Confidence 162
Time Limitations 166
The Knowledge Base of EAP 167
Chapter Summary and Practical Lessons Learned: A Re-positioning 173

7 **Language Across the Curriculum** 176
Bridging the Content/Language Divide: A Heuristic 177
Collaboration and Co-construction 180
Developing Pedagogical Content Knowledge 184
Chapter Summary and Recommendations 191

8 **Implications** 193
Policy and Strategy 194
Approaches and Practices for Teaching and Learning Within an English-Speaking Environment 196
Conclusions 200

Afterword: The Engaged Scholar 202
References 204
Index 216

Acknowledgements

There are many people and places that have contributed to and supported me in writing this book in a number of ways. By listing the places and organisations, I hope the people who are an integral part of each of them will recognise the part they played in helping me to complete this project. So, I would like to thank: the Leeds Institute for Teaching Excellence (LITE) for the opportunity of time and space to conduct the project discussed within this book; the Language Centre and BALEAP for providing me with a community that is both supportive and challenging; Bradford Synchronised Ice Skating Academy for the much needed escape!

There are, though, some key people to whom I would like to give specific acknowledgment. Crucially, thank you to all of those who gave me their time, thoughts and experiences by participating in my project. I hope I have represented you all fairly. Alex Ding has listened patiently, mentored wisely, given constructive feedback and useful insights as well as providing ideas, including titles when I got stuck! Melinda Whong's leadership created the space, push and confidence I needed to get started. I would also like to thank Ian Bruce and Cynthia White for their advice, support and feedback throughout.

My final thanks are for my family. Denise, my mum, is the best critical friend I could ask for, and my dad, Michael, has listened patiently throughout. My husband and children, Harry, Fred and Edie have tried very hard to be interested but generally cannot really see the point. In this way all three of them have shown great love while maintaining my sense of perspective. Thank you!

Preface

This book focuses on the nexus of language, disciplinary content and knowledge communication specifically at taught post-graduate (TPG) or Masters level. It aims to consider this nexus from the perspective of the multiple actors who are enmeshed in the consequences of the economic, cultural and ideological forces of Higher Education's current push for internationalisation. In addressing this interplay, I suggest the need for a greater synergy between language and content experts. I also suggest that change needs to be implemented through policy rather than on an *ad hoc* basis by individual teachers and that this involves a change to institutional educational and academic cultures. Therefore, it is a call to action for English for Academic practitioners to find a way out of the silo of their own centres and work to assert influence over the wider context in which they work.

The book emerged as the result of a scholarship project that was funded by the Leeds Institute for Teaching Excellence (LITE). The focus of the project was around exploring the significant roles language plays in shaping discipline specific knowledge and understanding. The specific 'research' question then became: *How do taught postgraduate tutors and students experience the intersection of language and disciplinary knowledge communication. What impact does this have on their identity?*

As an English for Academic Purposes (EAP) practitioner 'operating on the edge of Academia' (Ding & Bruce, 2017) the time and space to engage in scholarship around teaching and learning afforded by a Fellowship in LITE was an unprecedented opportunity. It enabled me to investigate questions that had troubled both myself and my colleagues for several years and to move from a place of guessing and knowing experientially to being able to gather evidence and a deeper understanding of what EAP should, or at least could, look like within my own context. It also placed me in the often uncomfortable position of being a somewhat accidental scholar, finding myself drawn into a different field of academic endeavour whilst trying to maintain my own strong identity as an EAP teacher.

This is a book that begins and ends, therefore, in practice. The focus throughout is on the practice of teaching and learning, on understanding the barriers and enablers to that practice within a particular context. Any theorising around this works to feed into future practice within the same context but can also be extended out from the specific to the general. As such, the outline of the project and introduction of the case study and participants comes prior to any theoretical review. I draw on multiple theories to then help explain the themes I see arising from the data and understand how this might be usefully used to impact on future teaching and learning practices. EAP practice, by its very nature, crosses numerous disciplinary and theoretical thresholds, depending to some extent on the students we are working with. By necessity, then, the EAP practitioner needs to have a broad working knowledge of a range of epistemological and ontological paradigms. Some of this is to enhance, develop and explain our own practice; some is to enable us to work within and across the disciplines of our students with a degree of confidence. This is what makes our work both challenging but endlessly fascinating. It is what makes our work shift from an epistemological to a praxiographic reality in which our knowledge becomes practically enacted in our interactions with students and colleagues.

As an EAP practitioner, the main theories I draw on as I work to understand the role that language plays in the taught postgraduate curriculum are from the broad field of language teaching and Applied Linguistics. I thus view some of the data through the lens of genre and discourse analysis, of Academic Literacies and Critical EAP. However, I extend beyond EAP, and consider whether the issues raised can be understood in terms of Threshold Concepts (Meyer & Land, 2003, 2005) and of curriculum design more broadly. I also draw on sociological theories of structure, identity and agency, considering the power structures and their impact on the social, cultural and intellectual spheres through a Critical Realist lens. In this way I hope to establish a symbiotic relationship between practice and theory that provides explanatory power rather than working to frame evidence within one theoretical framework and then suggest the implications for practice.

By making this project public, I am aiming to meet Shulman's definition of scholarship of learning and teaching that we

> develop a scholarship of teaching when our work as teachers becomes public, peer-reviewed and critiqued. And exchanged with members of our professional communities so they, in turn, can build on our work. (Shulman, 2000: 49)

While the project I draw on for this book was on a larger scale than many other scholarship of teaching and learning projects, which tend to focus on questions local to an individual practitioner's teaching, I am not making any claims that the experiences I outline or conclusions I draw here are fully generalisable beyond the participants and situations involved. However, I do hope that readers will be able to relate to and recognise similar experiences and patterns within their own educational context, and then make slightly better-informed decisions as to what to do in their own teaching and learning practices and that these local experiences will have global resonance. I also hope that those involved in curricular and policy decisions might also be able to draw some value from the heuristic for a language embedded curriculum that I propose

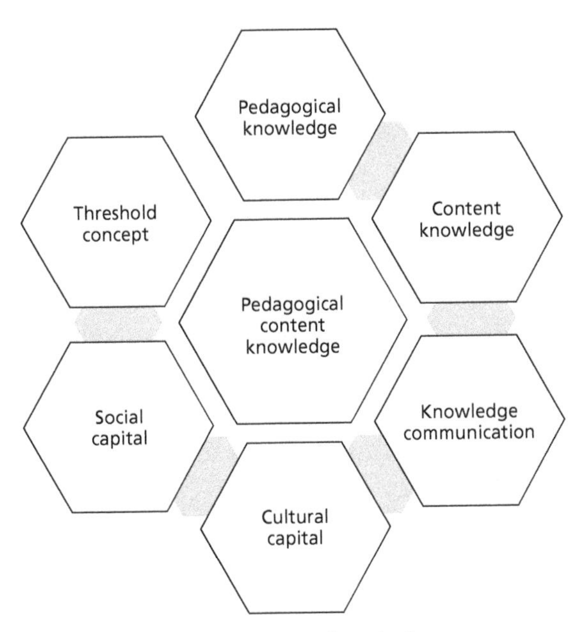

Figure 1 A heuristic for a language connected curriculum

(see Figure 1 here and discussed in detail in Chapter 7). This heuristic also provides some organisational logic for the rest of the book.

Organisation of the Book

This book is primarily for EAP practitioners. However, it is also aimed at others who work in Higher Education and have an interest in teaching and supporting international students. As it focuses on the taught post-graduate curriculum within the United Kingdom, it will be of most obvious relevance to those working within this context and teaching that level of students. However, my intention is that many of the questions raised and issues discussed will resonate with and can be applied to those teaching in other contexts and at other levels of study. At the end of most chapters, as part of the conclusion, I include a short summary that draws out the 'practical lessons learned' from the ground covered within each chapter. My final chapter also focuses on recommendations for policy and practice development at both institutional and local level. These, I suggest, are where EAP practice should position itself and work with university leadership to effect real change to teaching and learning in Higher Education.

The introduction to this book defines the conceptual and contextual parameters around which the rest of the book is built – considering what is meant by inclusive education and internationalisation in Higher Education, and how the work of English for Academic Purposes currently sits within this context. I then aim to provide a purposeful progression from specific to general; local to global throughout the rest of the book. In this way my own contextualised project becomes relevant to wider Higher Education practices.

Chapter 1 should be of interest to those who are developing their own scholarship project. As Ding and Bruce (2017) have already argued, scholarship is key to establishing a more equal status for EAP within the academy. In this Chapter I describe my own accidental and messy journey into the Scholarship of Learning and Teaching (SoTL), positioning myself and my work within a developing definition of SoTL. I focus on the complex ethical considerations that a SoTL project requires as well as the professional difficulties and benefits and institutional gains that arise from SoTL work. The methodological process is also described, positioning the project and the participants within a Critical Realist paradigm of methodological pluralism (Porpora, 2015).

The following four chapters provide a rich and complex portrait of the interweaving factors and themes that rhizomatically or interconnectedly create the experience of those involved in learning and teaching on a taught post-graduate programme. Each section begins with a brief focus on an (eclectic) theoretical theme, which then provides the basis for analysis of the data, practice and literature in order to demonstrate how they are consecutively interconnected and disparate. Building on expressions of identity, agency, temporality and trust, I consider how language is then perceived, understood, learned and taught as part of the curriculum, drawing on a range of theories and concepts to describe a structure that is undergoing elaboration as a diversified population interact within it.

In Chapter 6 I move towards a focus on English for Academic Purposes (EAP) and the role that EAP practitioners currently, and possibly could, play in supporting a more language aware TPG curriculum.

In Chapter 7 I present a heuristic that draws together the rhizomatic threads and can be used when considering where language might impact on the TPG experience for all those involved.

Finally, I suggest that it is only through strategic change and institutional policy that is then supported in practice that the required change to conceptions of language within the content curriculum can take place. I outline the various aspects of policy that need to be considered and make some suggestions for developing local practice in order to enable the development of a more inclusive, language-connected curriculum. It is here that EAP practitioners really need to begin to move beyond their teaching centres and establish networks and spheres of influence; in this way they will go further towards 'shortening the gap between what (EAP) is and what ought to be' (Ding, 2016: 13)

Introduction: Contextualising the Problem, Defining Terms

In this Chapter I aim to provide a contextual backdrop to main the focus of this book, that is how language and disciplinary knowledge are perceived as intersecting and disconnecting within a taught post-graduate curriculum and how EAP practitioners can work, therefore, to make language more visible across the University. I outline in broad terms the concepts and global themes of inclusion, internationalisation and English for Academic Purposes, considering how they interplay across the Higher Education landscape and create the structural conditions and backdrop that led to the key issues I aim to address. In doing this I explicitly and deliberately position the local and contextualised scholarship project and findings within a much broader, global conversation that is relevant not only for Higher Education institutions within the 'inner circle' countries but for all institutions that have 'internationalisation' as part of their core strategy, particularly when part of this strategy demands elements of English as a medium of instruction.

Aims and Purposes

Macro level discourse around finances and student recruitment have a direct effect on the micro level of the classroom. It is here that the real impact of a university's internationalisation policy is felt, as teachers and students need to learn, but often fail, to work together across and between cultures, languages and educational backgrounds. This book, then, aims to address the questions raised by this need in terms of student education, focusing on the nexus of language, disciplinary content and knowledge communication specifically at taught post-graduate (TPG) level. In doing so, I touch on key issues of internationalisation, inclusion and of teaching excellence in Higher Education.

I position language as central to all three concepts and argue that the teaching and learning of English for Academic Purposes (EAP) can be both a driver of and a solution to many of the questions arising from both internationalisation and inclusion agendas. In doing this, then, it is necessary to outline and problematise the label 'international student' and consider what issues are highlighted through working with these students, and whether they are any different to those raised by working with other student demographics. The drive for internationalisation should not be considered in isolation of a wider shift in approaches to higher education, and who should and does have access. The growth of an international student body (itself nothing new) has coincided with the push for access for all and an emphasis on social justice and inclusion. I therefore also consider internationalisation within a framework of inclusive education, outlining the current state of the debate here. It is this context we must begin with.

What is Inclusive Education?

In 2017 the UK government's Department for Education provided guidance to the Higher Education sector to support them in 'expanding their inclusive teaching and learning practice' as a way of enacting the 'Government's social mobility agenda – [of] giving everyone, regardless of their background or circumstances, the chance to study at higher levels of education' (Universities UK, 2017: 2). One of the main drivers behind this was the UK Government's 2010 Equality Act which required 'reasonable adjustments' to be made to ensure those with disabilities were able to access education without discrimination. Whilst the guidance was produced by the Disabled Student Sector Leadership Group (DSLG), inclusive education is now understood to encompass far more than a consideration of what adjustments could be made to 'usual' teaching practices in order to accommodate those with a disability. Inclusive learning and teaching is a much broader concept, and requires a consideration of a diverse range of needs and potential barriers to accessing the curriculum that could be encountered by any student so that adjustments for difference do not need to be made. In this way, the inclusion agenda in Higher Education encompasses other access campaigns such as Widening Participation (WP), decolonising the curriculum, a focus on first generation students, as well as a concern for those with differing physical and learning abilities and mental health issues. Inclusion is thus defined as 'issues relating to *all* students and to types of teaching and learning that fully and equitably include everyone in the classroom or in the programme cohort' (Grace & Gravestock, 2009: 1) and 'refers to the ways in which pedagogy, curricula and assessment are designed to engage students in learning that is meaningful, relevant, and accessible to all' (Hockings, 2010: 1) Ultimately, inclusive learning and teaching acknowledges that the (somewhat mythical) 'traditional' student should not be the held up as the norm, and that consideration of diversity and of intersecting needs and differences should be made when planning a curriculum, programme or module, whilst also understanding that 'students don't want to stand out as different yet want to be recognised as individuals' (Hockings, 2010). The drive for an inclusive curriculum must, therefore, include consideration of the similarly complex and intersecting needs of international students. In order to do this, however, it is necessary to understand, or at least problematise, what is meant by the term 'internationalisation' and 'international student'.

Internationalisation as a Driver for Change

Internationalisation has become increasingly prominent in Higher Education (HE) over the last decade. It is now deeply embedded in the structure and strategies of most Higher Education Institutions (HEIs); universities, at least in the UK, have become heavily reliant on the income brought by international student fees to support and maintain the institution. Many, if not most, HEIs have a Pro-Vice Chancellor (or equivalent) for Internationalisation, an International Office and an Internationalisation strategy which largely focuses on increasing international student recruitment and developing international research partnerships. While internationalisation in HE is seen as 'the integration of an international or intercultural dimension into the tripartite mission of teaching, research and service functions of Higher Education' (Maringe & Foskett, 2010 in Jenkins, 2013: 2–3), its main focus remains on developing a campus with a large number of international students, rather than on developing a truly international culture that is embedded across all HE practices.

Concurrently, the media focus on the internationalisation of UK campuses can be broadly separated into two strands. The more negative reporting around international students highlights individual institutional stories of issues with academic integrity, including plagiarism and contract cheating. Students from outside the United Kingdom are a frequent focus of the blame for increasing cases of fraudulent academic practice and the dumbing down of education (see, for example, The Guardian, 2019b; The Times Higher Education, 2019). Within these stories there is little nuance; 'International' students are represented as a problem.

When positive, the media focus is on the huge wealth these students bring into the country. In 2011, international students brought £10.2 billion in fees and spending to the United Kingdom (HM Government, 2013). Within the Higher Education press, these figures feed into the recurring debate with government around whether international students should be included in immigration figures or given visa extensions post-study. Most of this debate seems to focus largely on the financial gains brought by the higher fees paid by these students and their spending power contributing to the national economy (see Adams, 2017 and Letters to Guardian Education, December 2016 for examples) and recent government policy echoes this focus. The picture thus created is of a choice being made in Higher Education where financial gain takes precedence over quality, integrity and academic rigour. Within all of this, there appears to be little focus on the cultural and knowledge gains of having an international student body; there is even less media attention paid to how we can work to fully engage and collaborate with these students (notable exceptions being Bothwell, 2017; Cooper, 2017; Mora, 2017; Moran, 2017), building a real sense of reciprocity and achieving the aim of showing 'commitment to international solidarity, human security … [helping] to build a climate of global peace' (Fielden 2011, in Margolis 2016: 52).

Thus, the media image is in sharp contrast to the ideal of a global campus presented on most University websites, where a range of cultures and languages come together, either physically or virtually, to share ideas and to learn and conduct research together. Given the increasingly globalised and interconnected world outside HE, is difficult to argue with the ideal of an institution that reflects this representation and works to prepare its students for success within a globalised economy. Thus, ideals of knowledge exchange develop an '*elective affinity*' (Zepke, 2015) with the more cynically financial push to increase international student numbers and benefit from their higher fee-paying structures. This 'elective affinity' is the essence of neoliberal policy making that EAP practitioners and international students increasingly find themselves at the sharp end of, arguably being seen as the physical embodiment of the marketisation of higher education across the globe.

As the drive to internationalise increases, it is necessary to stop and question which of the pictures described above (if any) is the current reality, and to highlight how the shift is impacting both students and staff as they work and study together.

Who is an International Student?

There is a large body of literature debating the terminology and implications behind the label 'international student' (Baker, 2016; Carroll, 2015; Margolis, 2016; Montgomery, 2010; Ryan, 2011). Many are now arguing that any student studying at tertiary level could and should be viewed and view themselves as an international or a global student (Leaske, 2013; Jenkins, 2013). Recent studies indicating the learning gain and increased employability attached to a period of study abroad (Universities UK, 2017) have added weight to the arguments around the concept of 'internationalisation at home' (see Beelan & Jones, 2015; Leaske, 2013), where opportunities for an international experience are provided to all students. From this perspective, all students, whether at university in their country of birth/citizenship or not can and should see themselves as international. Here, being an international student provides opportunity that is not currently available to all; the aim to open up the opportunity of global mobility to others is based within concepts of widening participation and education for social justice.

However, within the UK HE context, the label 'international student' is used institutionally as a financial differentiation, denoting those students who are expected to pay higher fees because they hold a passport from a country outside the EU. University websites include pages specifically for 'International' students and include advice on, for example, visa applications, police registration and qualification equivalencies. In this sense, the label is simply of administrative use, allowing institutions to signpost those who need it towards the relevant information necessary to allow them to gain access to their chosen site of education.

Although the administrative differentiation, working along financial lines, separates students into groups based on whether they are 'Home', 'EU' or 'International', these lines become blurred and almost irrelevant once learning begins. This differentiation does not, crucially, neatly separate those students for whom English is an additional language from those who use English as their dominant or only language. It also does not separate those who have studied within an educational culture which is the same or similar to that of their current University. It is at this point that the label 'international' when applied to students studying in the United Kingdom can take on multiple shades of meaning. At times it is used to describe any student coming from outside the UK, but more commonly it is used as shorthand for any student who is studying in English when English is an additional language. At this point, the labels 'home' and 'international' become interchangeable with native (NS) and non-native speaker (NNS).

However, the terms native and non-native speaker have been also widely critiqued within linguistic research literature (Kachru, 1982; Holliday, 2010, 2011; Seidlhofer, 2011) and do not provide a clear distinction between those students who, for example, speak one language at home (their 'mother tongue'), yet have been educated in a second language and are proficient in both – often with greater expertise in writing their 'second' language as it tends to be the language of their education. When simplistically understood, the terms can, at their worst, be racist. At its best, the 'commonsense view' (Davies, 2003: 24) of a native speaker does not incorporate or question the multiple terms that should be implied, including the multiplicity of Englishes used across the globe ranging from English as a Lingua Franca (ELF) which involves communication in English that is negotiated between individuals who are all 'non-native' speakers of the language, to the concepts of bilingual codeswitching, translanguaging or even to regional dialect (see Canagarajah, 2013; Jenkins, 2013, 2015; Mauranen, 2012; Seidlhofer, 2011). All of these 'Englishes', whilst valid, may contribute to a student encountering difficulty in accessing certain elements of a UK university curriculum (Lillis *et al.*, 2015; Ivanic, 1998). There have been some recent empirical and pedagogically oriented studies on ELF and translingualism (for example, Flowerdew, 2015; McIntosh *et al.* 2017), although there remains disagreement as to whether this is simply a reframing that ignores prior research findings (see Matsuda, 2014; Tardy, 2017 and Tribble, 2017). Davies (2003: 8) suggests that native-speaker membership is one of 'self-ascription not of something being given' and is largely a sociolinguistic construct relating to levels of confidence and identity. It is thus, he suggests, a boundary that is 'as much created by non-native speakers as by native speakers themselves' (2003: 9). However it is understood, the binary use of the native and non-native speaker label is clearly as problematic and contested (if not more so) as the term international student.

Thus, when using the term 'International student' it is important to recognise the power structures and cultural capital (both in terms of opportunities and prejudices) that lie behind it. It is not benign and can be used to separate out and 'other' specific groups of people as well as to provide access to support. There is also no one label that can be used for this group of students that is not seen as denoting some kind of deficit differential, as any label must by its very nature be seen to separate one group from another, and measure one group against what is currently accepted as a standard norm.

There is not, therefore, one term which succinctly defines the students that this book is mainly concerned with, other than to suggest that at one time or another it is likely to relate to *all* students regardless of their nominal linguistic background. However, for the sake of ease, I will use the terms *International student* and, more frequently, *EAL (English as an additional language) student*[1] to denote those students who have traditionally accessed English for Academic Purposes classes and whose difficulty in accessing or voicing their understanding of the knowledge base of their discipline is more likely to be perceived as being due to their English language proficiency. Most commonly, but not always, these students enter University in the United Kingdom with an IELTS[2] level of 6.5 overall, or B2 on the CEFR[3].

Whilst these students have diverse profiles, motivations and needs, it is that they are using English as an additional language, as a medium of instruction, for academic purposes, that defines them as a distinct group. It is this language use that is perceived by the students themselves and the staff who work with them as the main barrier to being able to access their education. In most UK HE institutions, the work to reduce or bridge this barrier is done by English for Academic Purposes practitioners.

The Position of English for Academic Purposes

English for Academic Purposes (herein EAP) is widely defined as 'the teaching of English with the specific aim of helping learners to study, conduct research or teach in that language' (Flowerdew & Peacock, 2001a: 8) and more recently as the means of giving 'students access to ways of knowing: to the discourses which have emerged to represent events, ideas and observations in the academy' (Hyland, 2018: 390). In theory, then, EAP practitioners work to enable international, EAL students and, less frequently, staff to access the content of their disciplines and bridge the language gap that is perceived to be the main cause of academic and disciplinary exclusion for this group of students.

However, while there is little disagreement over what EAP teaching *is* (or at least should be), there is less consensus over *how*, *when* and *where* EAP teaching should take place, or indeed *who* an EAP practitioner should be. Tribble, for example, has suggested that 'accounts of what is meant by EAP' are 'fragmented and sometimes contradictory' (2009: 400); Ding and Bruce (2017) have provided a comprehensive overview of the marginal position that EAP practitioners currently occupy within University structures and suggest that this lack of status is, to a degree, self-inflicted. While there is a broad knowledge base for EAP to draw on – which Ding and Bruce suggest specifically are the research areas of: Systemic Functional Linguistics (SFL); genre theory; corpus linguistics; Academic Literacies and Critical EAP (2017: 66) – they also argue that this knowledge base is under-explored by teachers and the divide between research and practice in EAP has reached a critical point.

EAP practitioners, according to Ding and Bruce (2017), by not getting truly involved in scholarship find that they are not viewed as an integral part of the academy (and often don't view themselves as such). In this way, EAP units open themselves up to threats from outside, private providers and a de-valuing of the work they do. This occupancy of the margins of the academic space also often leads to a physical and structural confusion around the place and value of EAP. While many EAP units, when not outsourced to private companies such as Into or Kaplan, are housed in 'Language Centres', these centres themselves are housed in a variety of University spaces – or 'third spaces' as Hadley suggests (2015). In the UK, these include being part of an Academic Development Unit, a separate service unit, part of central student services; a 'wing' of an Academic School (usually Education or Languages and Cultures, but occasionally in less obvious places like Business Schools) as well as being fully integrated into an Applied Linguistics department. Depending on the positioning within the structure of the University, an EAP practitioner will have greater or lesser impetus, time and resources to engage in developing a knowledge base that takes them beyond the delivery of provided EAP materials and working to bridge the research practice divide.

This position is exacerbated by two further external influences. The first is the nature and timing of the EAP teaching year. Financially, the most lucrative teaching period for EAP is the summer, when international students arrive to take a pre-sessional programme for, commonly, 6 to 12 weeks prior to joining their academic programme in mid- to late September. The purpose of these pre-sessional programmes is to prepare international EAL students for the linguistic and literacy demands of university study in English; in the main however, the students who attend pre-sessionals do so because they have not yet met the language proficiency requirement of their academic programme via an IELTS or equivalent test score and are able to use pre-sessional assessments as an alternative. Pre-sessionals are intense periods of teaching and learning, involving both an exponential increase in student numbers for the EAP unit, and a commensurate increase in staff to teach them. This means either that an EAP teacher is engaged to work on short-term contracts and has limited job security and is thus denied the financial and institutional resources and support to engage in activities that enable them to develop and deepen their knowledge base, or, for those fortunate to have more permanent contracts, the traditional time for reflection on practice and for scholarship that is available to others on teaching and scholarship contracts, is not available. This context perpetuates the position of the EAP practitioner as just 'a "language teacher" with no connection to political and social issues' (Gee, 1990 in Turner, 2004: 107).

The second, powerful external force, is the relative hegemony of the IELTS exam as an indicator of language proficiency for international EAL student entry into UK education. Although in some contexts, Universities are beginning to accept other measures of language and academic literacy skills, EAP practitioners frequently find themselves having to build their teaching around the impact of this exam – whether it be moving students away from habits developed as a result of studying for the test (which, for example, requires students to write only 250 words of unreferenced argument in response to a generic 'essay' question), or helping students prepare for the test itself. EAP teaching and IELTS teaching are often wrongly conflated. Outside the EAP and language teaching community, there is only a vague understanding of what the different IELTS levels mean (Benzie, 2010; Murray, 2016a). Thus, academics and students alike can often work on the assumption that the stated entry level for TPG study (typically in the United Kingdom 6.5 overall with no less than 6.0 in any of the four skills of reading, writing, speaking and listening) equates to a level that will enable access to the study content with no further need for language development. In fact, contrary to this belief, IELTS itself suggests that at 6.5, for a 'linguistically demanding academic course' that 'English study is needed'; for 'linguistically less demanding academic courses', the student's level of English is 'probably acceptable' (IELTS, 2017). This does beg the question as to what kind of TPG programme could be classified as linguistically less demanding, particularly if it is one studied in the United Kingdom, alongside a global community of peers for whom English is the only common language for knowledge and social exchange.

This conflation between IELTS and EAP places EAP practitioners in a difficult position because, arguably, it devalues the complexity of the work involved in de-coding disciplinary knowledge communication discourse and in working with students to enable them to access the academy. Critically, it is often the EAP unit that becomes the target of blame when students on a programme are deemed to be struggling due to language proficiency. Thus IELTS and concurrently language is viewed as the 'catch-all term for problems with unmet standards, and the need for remediation' (Turner, 2004: 99) which Turner also argues results in the denial of academic respect to EAP teachers and their students. 'The dilemma for the academic literacy pedagogies is that they are only tolerated while they remain remedial … the remedial positioning of language work is necessary in order to maintain the culturally embedded and socially embodied "habitus" of being academic' (Turner, 2011: 37), in other words language is seen as part of the physical embodiment of an academic; it is 'who they are' or 'who you become' implicitly rather than something that can be analysed, de- and then re-constructed explicitly and expertly. I argue that it is part of the role of EAP practitioners to change this perspective amongst their academic colleagues that few are, as yet, fulfilling.

Thus, while many international students, are struggling to access their own academic 'community of practice' (Lave & Wenger, 1991) because the current apprenticeship does not explicitly acknowledge its shared language as something to be learned and are thus facing perceptions of being in deficit linguistically and feeling culturally excluded, their first point of contact is often with EAP teachers who are also relative outsiders to the academy. Many of these practitioners, particularly over the summer pre-sessional period, are employed with little experience, little time to equip themselves and occasionally dubious credentials. This does, then, raise the question as to how well EAP does the job it is tasked with doing, of giving 'students access to ways of knowing' (Hyland, 2018: 390). If EAP units, and language learning and teaching, are disconnected from and undervalued by, the rest of the academy, how do students, EAP teachers and content teachers understand where language and content knowledge connect and disconnect? How does this view of language and its place in knowledge communication impact on their teaching and learning practices and their identity as members of an academic community? And how can teaching and learning be truly inclusive and international if a focus on the (globally dominant English) language used to communicate the knowledge being gained is outwith the written curriculum? Within this book, I aim to draw out these complex themes and questions, demonstrating how they overlap, intersect and, at times, contradict. In doing so, the narratives and experiences I present feed into a global conversation and hopefully present some clear suggestions as to how to think differently and work more collaboratively to ensure that language becomes more visible across the higher education curriculum and that all students are better supported in accessing and demonstrating their own emerging knowledge.

1 The Accidental Scholar

While not the primary focus of this book, I feel it is necessary to outline my own position in relation to the investigation I undertook, and the context within which it took place. This is, in part, to acknowledge the contextualised and subjective nature of the study. As a participant as well as investigator in an ethnographic study this needs to be highlighted and recognised (Street, 1995: 51) whilst maintaining its relevance beyond the local. I also hope to provide some insights into the messy process of scholarship of learning and teaching, how it fits into the wider Higher Education landscape in the United Kingdom and to encourage other EAP practitioners who are considering or developing their own scholarship profiles.

The scholarship of learning and teaching has only recently become a focus of strategic attention, at least in the United Kingdom, in line, as I discuss later, with a greater focus on teaching as well as research excellence. This can be evidenced by the growth of centres for teaching excellence across the sector. These can take the form of either being centrally funded and cross-disciplinary or arranged around specific disciplinary concerns. For the increasing number of academic staff who are employed on teaching as opposed to research contracts engaging with this form of scholarship can be viewed as both an institutional expectation and a route to promotion. However, there can also be an assumption that those involved in the scholarship of learning and teaching in Higher Education will already know how to meet this expectation and that they have the skills required to undertake investigations that will provide insights into and enhance student education in an ethical, purposeful and rigorous manner. There is often little acknowledgement that for some, the move from disciplinary research into the scholarship of learning and teaching requires a complete epistemological shift and a different set of skills and dispositions, whilst others, employed for their professional practitioner expertise, have undergone little previous research training at all (Geertsema, 2016). For many of us, this move into scholarship is unplanned and accidental. In this chapter, I share my own rather haphazard journey into new territory in the hope that it might shed some light on the process for others who find themselves 'accidental scholars'.

I begin this chapter by defining the scholarship of learning and teaching as distinct from (educational) research and placing it in the current context of the need to provide metrics and quantify teaching excellence in HE. This contextualisation adds explanatory power to my own personal journey, which is key to the development and trajectory of this book. I then provide an outline of how the project that is the main focus of this book developed in terms of methodology, data collection and analysis and how I worked to both theorise this process whilst working to maintain practical relevance throughout.

The Scholarship of Learning and Teaching

The Scholarship of Learning and Teaching (henceforth SoTL) movement began in the United States and its origins are widely attributed to the work of Ernst Boyer (Fanghanel *et al.*, 2015). While SoTL remains 'a relatively ill-defined concept' (Fanghanel *et al.*, 2015: 6), there are points of agreement as to what SoTL should involve. Shulman's (2000) suggestion that scholarship should be made public and open to critique is now widely accepted and has since been built upon by Felten (2013) who suggests a 5-point framework for SoTL as 'inquiry focussed on student learning; grounded in context; methodologically sound; conducted in partnership with students; appropriately public'. Essentially, SoTL is the 'systematic study of teaching and learning, using established or validated criteria of scholarship, to understand how teaching (beliefs, behaviours, attitudes and values) can maximise learning, and/or develop a more accurate understanding of learning, resulting in products that are publicly shared for critique and use by an appropriate community' (Potter & Kustra, 2011: 2). While this is the 'norm', it is also possible to view SoTL as going beyond this instrumentally narrow relationship to teaching and learning, where teaching is viewed as the cause, leaving an effect on students[4]. Here, SoTL becomes a reflexive examination of our own beliefs and practices in a way that 'enables students' voices and perspectives to be fully integrated into not only in problem solving scholarship but also wider educational discussions concerning ideas, theory, values and purposes' (Ding, 2016: 15–16).

SoTL is alternatively referred to as pedagogical research and is largely seen as specific to those who work in Higher Education. It is distinct from, yet a branch of, educational research. The key differences being that educational research is conducted by those specifically research trained within the epistemological paradigms of educational research; it is often viewed as research *on* or *about* a particular area of inquiry. Those entering a SoTL 'community of practice' (Lave & Wenger, 1991) do so from a wide range of disciplinary epistemologies and are required to both maintain this particular perspective and cross the 'tribal' boundaries of disciplines (Becher & Trowler, 2001) to develop scholarship projects that can be accepted as a worthwhile contribution to teaching and learning inquiry; it is generally viewed as being research *for* a specific group or community. By doing so, scholars open themselves up to the question posed by Kanuka (2011 in Fanghanel *et al.*, 2015: 9): 'Notwithstanding such small-scale efforts [i.e. inquiry on and for practice] may make contributions to one's practices – but when they are made public, is this enough to be considered a scholarly contribution?' In their SoTL Manifesto, Ding *et al.* (2018) argue that SoTL should be about impact, including 'impact on people, policies and practices (assessments, concepts of syllabi and curricula, communities, community engagement, leadership, mentoring)'. By having impact beyond the individual teaching and learning context, they go on to suggest that 'scholarship has the potential to enable language educators to actively shape their educational contexts rather than be shaped by circumstance, others and powerful ideologies and structures' (2018: 58–59). This, I suggest, is the guiding principle of this book and should be a key guiding principle for all EAP scholarship and practice.

What counts as SoTL methods of inquiry is still an area for debate, and the lines between a pedagogical approach and a means of inquiry into teaching and learning can, at times, become blurred. SoTL is also tied in with continuous professional development, or learning (CPD/CPL) (Geertsema, 2016). SoTL should take place within disciplinary contexts but can also be used to cross disciplinary boundaries, encourage educational development across disciplines and faculties and work to change institutional practices. Viewed in this way, that EAP practitioners should engage in SoTL seems obvious. There has been long standing consideration in language teaching research around how/whether practitioners engage in research (Borg, 2009, 2013; Hanks, 2019; Smith & Rebolledo, 2018), with a focus on Action Research, Exploratory Practice or Exploratory Action Research (for more comprehensive discussion of these areas of language practitioner research see, *inter alia*, Burns, 2010; Allwright & Hanks, 2009; Smith & Rebolledo, 2018, respectively) . The imperative for language teachers to do this remains unclear throughout much of this literature, other than a sense that it is desirable and could, in the case of Exploratory Practice, enhance 'quality of life' (Hanks, 2017). For those working as EAP practitioners in Higher Education, the imperative goes beyond this because 'by withholding contributions to scholarship we are potentially limiting our own agency, limiting our ability to influence structural change and accepting of changes and practices defined and decided by others' (Ding, 2016: 12). SoTL in EAP, while more ambiguous in terms of methodology, becomes an 'attempt to shorten the gap between what is and what ought to be' (Ding, 2016: 13), i.e. moving EAP practice from a liminal, marginal position to one with academic status and a central place within the academy.

In their review of SoTL literature for the HEA, Faghanel *et al.* (2015) outline the different contractual status with which people working in academic departments now find themselves. Within this, they suggest that there is an increasingly clear split between those on research focused contracts, with a requirement to submit to the Research Excellence Framework (REF) and those on teaching only, or teaching and scholarship, contracts. One of the aims of SoTL is to redress the perceived imbalance in status between teaching and research in HE institutions, providing public evidence of excellence in learning and teaching and thus enabling those on teaching contracts to be rewarded and recognised in equal measure with those on more research focused contracts. This has now been extended to institutional level with the introduction of the TEF.

The Teaching Excellence Framework (TEF), introduced in 2016 aimed to place teaching in HE on a par with research. In the first round of TEF submissions, evidence was required at an institutional level only, with HEIs being able to create their own narrative around excellence in teaching and learning. Evidence of engagement in the scholarship of teaching and learning, or pedagogical research, was a key part of many institutions' submission. As Fanghanel *et al.* suggest (2015: 10):

SoTL has been utilised as:

- a means of **demonstrating excellence** with a view to raising the status of teaching in relation to that of research;
- a framework to **evidence excellence** in teaching and learning and **assess** teaching quality;
- a tool to **develop academics** and teaching **practice.**

In this way, SoTL is no longer only a movement of engaged practitioners moving towards evidence based teaching practices, working to legitimise, to share and go public with their investigations and do so in a rigorous and systematic manner. SoTL is now being used as evidence for individual promotion and also for evidence of institutional excellence.

There are, therefore, instrumental and external pushes for staff to engage with SoTL; UK HEIs are increasingly setting targets around how many of their teaching staff should have achieved Fellowship status through Advance HE, one requirement of which is engagement with SoTL literature and practice, and promotion criteria have been developed to encourage a route through student education and pedagogical research. Therefore, a number of academic teaching staff may find themselves being pushed into SoTL with little intrinsic motivation and understanding around how SoTL can feed into their student education practice and little sense of purpose or direction. Without this understanding, the usual considerations, particularly around ethics, that would take place when undertaking disciplinary research can sometimes be lost through a sense that SoTL is less rigorous, more personal and a part of 'normal' teaching practice.

The ethics of SoTL

The ethics of SoTL, as with SoTL itself, remain undefined and a somewhat grey area. By attempting to define itself as different to traditional research, and separate from Educational Research, it is tempting for scholars to proceed without the same rigorous consideration for ethics as is now required before conducting research. This is exacerbated because many SoTL projects are localised and, by definition, should be part of normal teaching and learning practices. However, as MacLean and Poole argue: 'Teachers who act also as scholars of teaching and learning in the practice of their discipline must consider the ethics of their **dual roles** in situations in which **their students are also their subjects of research**' (MacLean & Poole, 2010: 1).

Most exploration of the ethics of SoTL has focused on this dual role in relation to students. Martin (2013) emphasises the need to consider students as 'human subjects'. In the same article, Martin reproduces a statement of ethics for SoTL created and presented by Gurung *et al.* at the 2007 ISSOTL conference in Sydney, Australia. In this they outline three major principles. These are:

- Respect for Persons: Students (the research participants) should be treated with autonomy and must be free to decide whether or not to participate in a research study unless archival data are being used or if results are not to be presented publicly.
- Beneficence: Instructors (researchers) must recognise the need to 'maximise possible benefits and minimise possible harm'.
- Justice: Students (research participants) should be the people who most benefit from the research. It would be unethical to research a particular group in excess if that group is not the group that will benefit from the knowledge generated through the research.

(Gurung *et al.*, 2007 in Martin, 2013: 62–63).

The British Educational Research Association (BERA) have also produced a comprehensive guidance document around the ethics of educational research (2018). While wider ranging in scope, it does also encompass more localised practitioner research, and follows similar principles of respect for all involved. However, whilst comprehensive in areas to consider and principles to follow, it works on the assumption that those engaged in this type of research are epistemologically social scientists, and that their local ethics approval committee will also be accustomed to consideration of ethics from this standpoint. As Martin (2013) points out, this is not necessarily the case for those involved in SoTL, where their Faculty ethics committee may be more accustomed to consideration of, for example, medical ethical research principles.

Within SoTL and BERA's guidance, the main focus for potential benefit and harm is on the students as research participants. However, once a project extends beyond the researcher's own classroom, it is also necessary to pay attention to the potential benefit and harm to colleagues and an institution. As a colleague, it is important to consider the impact different relationships might have on the data gathered, and that comments from peer participants may be more unguarded than with an unknown researcher. In addressing an issue that is potentially problematic for many colleagues, a SoTL project is likely to result in some data that is difficult to report whilst remaining collegial and supportive. Here, I would argue, it is necessary to carefully follow the second of BERA's principles, that a researcher 'should respect the privacy, autonomy, diversity, values and dignity of individuals, groups and communities' (BERA, 2018: 4).

Most importantly, the ethical goal of any SoTL project, as with any other form of research, is 'to maximise benefit and minimise harm' for all involved. Within my own project, I always strove to work towards these principles. All participants were volunteers and were fully informed of the aims and purpose of the project. Written consent was obtained for use of all non-public documents (for example student work; class materials and assessments) as well as for the use of interview transcripts. On occasion, participants asked for certain comments to be 'off the record'; this request has been respected at all times. Finally, the interactions I had with participants were approached as opportunities for benefit to all, maintaining a sense of investigator-participant reciprocity.

There were multiple occasions, however, when I needed to wrestle with my conscience and question where my own ethics lay. It quickly became clear that my investigation was not benign; that there were a range of tensions and emotions at play. There were conflicts around whether my actions might harm students in protecting staff or *vice versa*. I hope I have navigated these tensions with sensitivity and given a representative voice to competing perspectives without causing undue harm in the process.

My journey into scholarship

I have provided this background context to situate myself within the wider UK higher education landscape. As an EAP practitioner, 'operating on the edge of academia' (Ding & Bruce, 2017), there is no one clear route into 'the academy', and scholarship or practitioner research, at least in terms of going public, has not ranked highly in the commitments of most practitioners to date. Reasons for this are myriad, but largely connected to teaching workload; qualifications; precarity and structural conditions (see Ding & Bruce, 2017; Hadley, 2015 for further discussion).

My own route into scholarship perhaps exemplifies this position, and maps onto the changing landscape of UK HE in terms of measures of teaching and excellence as outlined above.

I became an EAP practitioner in 2000, during the first real boom phase in international student recruitment to UK HEIs. I was recruited because of my qualifications and experience as an English language teacher, having worked for a number of years in private language schools in a variety of countries. These qualifications are typical of those requested for entry into teaching EAP – a Diploma in English Language Teaching (DELTA) – with little or no focus on EAP specifically. I was initially employed on an hourly paid contract. In order to qualify for a more permanent position, I studied for a post-graduate degree in language teaching. However, there was, and remains, no requirement to demonstrate expertise or understanding of EAP specifically. There is an assumption that this is something that is developed 'on the job' (see Ding & Campion, 2016; Campion, 2016).

Beyond completion of my Masters degree, my scholarship was desk based in terms of reading the research of others around the teaching and learning of EAP, and then attempting to apply this research to my own classroom practice. As my Centre grew in size, I was asked to take on programme leadership responsibilities, so was able to expand my understanding of EAP beyond my own classroom. From there I also developed an interest in supporting others in their own professional learning. Other than a few presentations at one-day conferences, the impact of my scholarship was internal and was largely entered into in order to prevent a personal feeling of becoming stale and stuck in the cycle of four-term, year-round teaching (Bond, 2017a).

In general, then, my journey into scholarship is the result of volunteering to take up new opportunities in order to prevent boredom, but without expecting to be successful. It is, in fact, one of constant surprise and of permanent imposter syndrome – of not feeling I was worthy or intellectually capable of being accepted or taken seriously within an academic context. This is both personal but also structural – as an EAP practitioner, there is very little precedence for being accepted by the wider academy. In fact, the mythology around not being able to gain access is close to doctrine. As John Swales wrote recently:

> Of course, we rarely have the time and the opportunity to be true ethnographers of researchers, disciplines or departments; perhaps the only people in our field with that kind of luxury would be those engaging in doctoral-level research or those fortunate enough to have generous sabbatical leave arrangements. (Swales, 2019: 11)

The suggestion here being that this is true of almost nobody in the field.

It was therefore by accident rather than design that I found myself in a position of being afforded the luxury described by Swales. The project within this book is the result of applying for the first round of fellowship support within the newly established Leeds Institute for Teaching Excellence. The most difficult question in the interview was 'how will doing this project impact on the next five years of your career?' This was almost impossible to answer as there was no parallel to measure against. By applying to take on this project, I found myself placed in the position of being an 'accidental scholar' with no real research training within any epistemological or ontological paradigm and, as an EAP practitioner, no real firmly agreed knowledge base upon which to build my project. The accidental nature of my scholarship is important to bear in mind as it defined, indeed dictated, the exploratory nature of the process I engaged in. It also explains the lack of one clear theory within which my work is located. Rather I became magpie like, borrowing theories from the disciplines I had contact with and through the discussions I had with students and colleagues, selecting when I felt resonance. My work was, and remains, incredibly 'messy' (Law, 2004). However, over time, I have begun to see my understanding as fitting broadly within a Critical Realist paradigm where there are no specific methods and the philosophy of which 'justifies methodological pluralism' (Porpora, 2015: 64). Both Critical Realism, SoTL and the broadly ethnographic approach I took demand that the reality of a local context is studied and that a voice is given to the participants involved. SoTL is generally understood as localised learning, aimed at development within that context. However, it is also a social movement aimed at transforming higher education when individuals talk about new things in new ways and others pay attention and learn from these stories, applying whatever they find as relevant to their own contexts.

The Project

Given my very clear identity as an EAP teacher and latterly scholar, as opposed to researcher, I preferred to frame my questions as 'project' questions rather than research questions. This connected to my sense that the investigation was one of reflexive, exploratory praxis. I was not aiming to find answers or solutions to problems, but rather to 'work towards understanding' (Allwright & Hanks, 2009) that would enhance my own student education practice. I began with the belief that this understanding may or may not lead to concrete recommendations beyond my own specific teaching context; if it did, the project would develop into a form of Action Research (Kemmis & Smith, 2008; Kemmis, 2009), where the impact of the recommendations would then need to be assessed and scrutinised. Whatever the outcome, it would undoubtedly lead to more questions.

However, working within a newly established centre for teaching excellence, as one of its first Fellows, I was also working within certain expectations and under a spotlight of institutional scrutiny. Whatever my own limited expectations of my work, the institutional ones were that I should feed into developments around institutional teaching practice and policy, make recommendations and suggest changes. SoTL is, after all, about impact. This institutional push felt, at times, to come too early and I was persuaded to go public via blogs and online videos before I felt I had something concrete to contribute. This external push did, however, also force me to formulate ideas where I might otherwise have procrastinated.

One such push was the need to succinctly explain the purpose of my project to others very early in the process. I initially outlined it as: 'understanding the significant roles language plays in shaping discipline specific knowledge and understanding'. The specific 'research' question then became: *How do taught post graduate tutors and students experience the intersection of language and disciplinary knowledge communication. What impact does this have on their identity?*

However, underlying this question were layers of other complex, inter-related questions that all impacted on my approach to the investigation as it developed. These included:

- Why do I need to understand this?
- Who else needs to understand the role language plays?
- How much more do I and these 'else' need to understand than is already known?
- How do 'we' currently understand the role language plays?
- How significant do we believe language is in shaping our disciplinary knowledge?
- What (kind of) language do we think is significant within our discipline?
- Do we think about language at all?
- Is language considered within teaching and learning practices?
- What does our understanding of language and a discipline say about the way the discipline is taught and learned (and assessed)?
- Is there consensus of understanding around the role of language: across disciplines; between teachers and students; between EAP teachers and content teachers? If not, is consensus necessary? If necessary, how can it be reached?
- What is the role of the student, the content teacher and the EAP teacher in developing an understanding of the discipline and its language?
- What do we mean by language?

Each one of these is in itself a complex question that continues to be debated in the field of applied linguistics and EAP. Language competencies alone are 'complex, dynamic and holistic' (The Douglas Fir Group, 2016: 26). I do not, then, attempt to provide comprehensive answers to any of these questions in isolation; rather it is necessary to highlight that a focused question created for one SoTL project does not emerge in isolation to others. The Douglas Fir Group have provided a useful framework that demonstrates the 'multifaceted nature of language learning and teaching' (2016: 24). If we are truly working to *understand* a real situation, we cannot construct a true picture if this complexity of intersecting issues is not considered.

Methodological Approach

As an 'accidental scholar' the theoretical and methodological influences on my thinking expanded as the project progressed. The aim was not to find a fit within a specific research paradigm or to find a disciplinary home. As an EAP practitioner it is necessary to have knowledge of wider academic communities in order to act as a bridge or an opener of thresholds for our students. As a practitioner it is necessary to know enough; engagement in scholarship forces a widening and deepening of your understanding and requires you to draw on this as you develop your thinking around practice. This is particularly the case when you make your work public and are required to conform to the social and cultural norms of academic knowledge communication. Throughout the process the sense that I was engaged in scholarship *for* practice and for practitioners and students, rather than research *about* practice remained at the forefront of my approach. The question 'How useful is (this) theory for practice?' was a constant, and frameworks or theoretical perspectives were chosen for their explanatory power in relation to practice within that context and time, rather than to provide theorised positions. This results in what could be viewed as a rather motley collection of theories, frameworks, perspectives and approaches.

First and foremost, it is important to recognise the influence I, as investigator, had in the collection of, as well as interpretation of, data. Working within the practitioner research paradigm, my approach was heavily influenced by Exploratory Practice (Allwright & Hanks, 2009) as this is where my journey into scholarship began (Hanks, 2015; Bond, 2017a, 2017b). I was not aiming to find clear answers to the questions I was asking, but simply to better understand the situation. I was working towards an understanding based on the principles of quality of life and collaboration that have been developed by Exploratory Practice research (Allright & Hanks, 2009; Hanks, 2017). As the principle investigator in this project, but also as a participant, the understandings I reached were intended to have a direct impact on my own student education practices as well as (hopefully) those of others. In fact, I viewed the heightening of awareness and the consequent co-construction of knowledge through the interaction between the researcher and interviewee (Guba & Lincoln, 1994) as one of the first means of having impact on the educational practices of my institution through this scholarship project. In this way I hoped to fulfil the SoTL requirement around the **'dissemination** of analyses of practice to inform others and developing intellectual **communities** and resource **commons**' (original bold script; Fanghanel *et al.*, 2015: 7).

As an investigation born out of practice and practice-based questions, the methodology used also emerged from this practice. It was by necessity rather than intentional design, ethnographical. As Swales has recently argued for EAP in general: 'We can and should aim for an insider "emic" approach, even if we cannot always achieve it, because the effort involved in trying to become something of an insider will often produce pedagogical and educational benefits' (Swales, 2019: 11). I was already deeply immersed in the context and institution under investigation and taking an ethnographic approach allowed me to recognise the intersubjective nature of the project I was undertaking. Ethnography, as 'participant observation', 'involving detailed descriptions of small groups and of their social and cultural patterns' (Street, 1995: 51) acknowledges my status as both investigator and participant. Academic Literacies (Lea & Street, 1998) provides a theoretical framework for ethnographic studies around language and literacies development, use and power and is also one of the knowledge bases that EAP draws on (Ding & Bruce, 2017). I therefore followed an Academic Literacies approach to data collection, which included observations of classroom interaction and analysis of student writing, as well as paying attention to the lives and roles of students and teachers outside the classroom. In this way, I built a detailed understanding of, and acknowledged the interaction between, classroom practice and the wider academy, and the role each plays in staff and students' academic, social and cultural experiences as each of these areas themselves impact on the others, adding to or detracting from the power of individuals and groups.

My approach also drew from other methodological constructs and theoretical perspectives. I draw parallels between the Academic Literacies concept of there being three 'levels' of approach to literacy development in Higher Education (skills development; socialisation and transformation) and the outlined phases of Threshold Concepts (Pre-threshold; liminal and over the Threshold, Meyer & Land, 2003). However, the overarching theoretical perspective I took was one of Critical and Social Realism, following Margaret Archer (1995), who argues that it is necessary to focus on the interplay between structure, culture and agency or between the individual and collective. This Critical Realist lens allows for a distinction to be made between what is observable and what is real; it accepts that all understanding is subjective. Archer emphasises the importance of temporality and argues strongly that society does not converge with the individual, or *vice versa*. Rather there is an interplay between a pre-existing structure and an agential self. She positions this as also developing through three distinct (although unpredictable and changeable) phases from structural conditioning to sociocultural interaction finally to structural elaboration. In this way a society or institution, although pre-existing, is formed and re-formed as a result of the individuals, roles, collectives who make up the human nature of that society. In this way it is important to study both the agential actors and the culture and society of which they are a part, building a picture and understanding from the interplay that takes place. This perspective also lends itself to an ethnographic case study approach (Porpora, 2015) as well as an Exploratory Practice paradigm where researchers work to understand rather than definitively answer.

I wanted to develop a detailed picture of education practices and linguistic understandings within different sites, with a different disciplinary focus and a different student population/cohort mix. I hoped to consider the significant differences in experience across sites as well as the similarities (Silverman, 2000). As 'using a case study provided a systematic way of examining language, identity and power' (Feagin *et al.*, 1991 in Sterzuk, 2015: 57), my ethnographical investigation developed into a multiple Case Study approach. Aimed at retaining 'the holistic and meaningful characteristics of real-life events' (Yin, 1994: 3) and allowing for a number of different sources and methods of collection to create a 'thickness' (Geertz, 1973) to the data, three sites within the larger institution being researched formed part of the project investigation. Within each of these Case Studies, the collection of data took on slightly different forms depending on circumstances (Yin, 1994). As with most ethnographic case studies, although the data collected can be categorised as qualitative rather than quantitative, there was not one specific form of data collected. Rather, I acted as curator of a collection of documents and conversations which could be seen to relate in some way to the focus of the project, noting comments and thoughts in a journal as the project progressed.

What ultimately developed from this rather messy, emergent research design was data that provides both individual narratives of a moment in time and around a specific experience and then a layered, deep, often contradictory and complex picture of an institution in flux as it interacts with different agents.

Project Design and Data Collection

As the methodological approach to the project being reported in this book was emergent and reflexive, crossing boundaries as my understanding of the issues and questions being investigated developed, the project design needed to reflect this exploratory approach. SoTL also requires that I acknowledge my disciplinary epistemologies. As an EAP and language teacher, my overall perspective of the University is subjective (Denzin & Lincoln, 2005). The perspective taken is of the entire University as a language classroom; a site of linguistic struggle, in both its literal and theoretical senses. This struggle is enacted within and through content knowledge, educational cultures and individual and collective identities, all of which directly impact on teaching and learning. All forms of communication involve language (if viewed in multimodal terms) in some form; understanding how language creates and develops knowledge and thought within a discipline is essential to the work of a University, and therefore the work of all those teaching in it. The main focus of this project is to consider how this language is used to express and connect ideas and how those working in the University engage with and understand its power.

Given this perspective, it was important to collect data that included all involved in the teaching and learning process – both EAP teachers, other teaching staff and students. It was also important to look at the influence of interactions and experiences that took place both inside and outside the traditional learning spaces, so across the entire TPG curriculum. The curriculum in this project being conceived as in its widest sense as 'the interplay of all those involved' in Higher Education and as a 'cultural imperative' (Barnett & Coate, 2005: 159).

The collection of data, then, took place across the three separate Case Study sites within my own institution. One of these was the EAP teaching unit, the other was an Arts, Humanities and Cultures (AHC) School and the other a Science, Technology, Engineering and Maths (STEM) School. This allowed for comparison across disciplines and for consideration of any (dis)connections between those whose primary role is to focus on language (the EAP unit); those who are concerned with content knowledge; and those (the students) who find themselves grappling with both simultaneously. The experiences and thoughts of both student and teacher were included, the aim being to create a sense of where and if overlap occurred.

The collection of data ran over a period of 10 months, from May 2016 through to March 2017.

Interviews and focus groups

The interviews and focus groups with teachers were all semi-structured and were built around a similar set of questions. The questions were designed to develop an understanding of how the teachers approached teaching at TPG level specifically, and how they viewed it as different to teaching at undergraduate level, before moving on to consider how they felt language had an impact on their disciplinary content teaching and assessment practices. I also asked them to consider any differences they perceived in their approach to international and home students and asked them what, if any, training they felt was necessary to support teachers when working with increasingly diverse groups of students.

EAP teachers were asked similar questions. The focus on content and language was reversed, asking how teaching EAP with a focus on disciplinary content impacted their language teaching.

Similar questions were used for individual interview and focus groups for the students, asking why they had chosen to study their TPG programme, what the differences were between their undergraduate programme and their post-graduate programme, and if they felt that language had any impact on their ability to understand and study on their programme. Questions also covered understandings around assessment practices, and what support they received and would like to receive from the University.

The student Focus Groups were re-interviewed on three occasions over the 10-month period. The structure of these interviews became looser throughout. The opening request 'Tell me how things are going', with guidance to think about their language use was usually enough. Having this more longitudinal element to the data collection, with multiple points of data generation, allowed comparison across the different data sets (Cohen *et al.*, 2007). Importantly, it provided insight into the development and change in the students' understanding of and approach to language, learning and content communication over time and across two sites, as these students initially began their time at university as EAP students before moving into their academic School.

Observations

By allowing direct access to events and interactions (Simpson & Tuson, 2003), observations allowed me to look directly at the interaction between language and content as it played out in the different learning environments of each of the three Case Study sites, rather than relying only on second-hand recounts (Cohen *et al.*, 2007). As the intended focus of the observations was the language/content interplay, I chose to use the usual observation form used for peer and evaluative observations of EAP teaching within the EAP unit. I chose this form because of my own familiarity with it, but also because it guides the observer to direct their focus on academic language and discourse. However, whilst fit for purpose in an EAP classroom, and useable in more discursive seminars, I found this format unusable in a practical or lecture session. Here the pedagogy and approach were different, rendering many of the observation questions irrelevant. In these sessions, other than noting down the arrangement of the room, I wrote my own field notes as a train of events, with some verbatim interactions. Much of the final product tended to be lists of vocabulary that were either totally unfamiliar to me as a non-disciplinary expert, or that I felt would not be taught with that particular orientation of meaning in a general EAP or language classroom. This in itself provided useful insight into the disconnect between a students' likely previous English language learning experiences and their real disciplinary experience of language in use.

Field notes and informal discussion

Here I distinguish between the notes taken in the more unnatural, and therefore more formal, situation of classroom observations where, however familiar the context, I was always an 'outsider', and the messier and more personal notes taken after critical incidents, to enable the development of thought processes or as an aide memoire.

The majority of my field notes were taken during the period leading up to, during and directly after the summer EAP pre-sessional programme that took place during the period of data collection. This was when I was most directly involved as a participant as well as an observer. Immersed as I was in interactions that directly addressed the questions I was hoping to explore, the daily interactions with students and teachers all held potential meaning. Taking notes enabled me as much as possible to view events and conversations from a different perspective, making the familiar strange and allowing myself to create the critical distance needed to begin to view the data as an outsider.

Beyond the summer, the notes taken became more erratic, and centred on critical incidents or more personal thoughts as I reviewed data previously collected.

My notes were generally divided between a detailed description of the conversation or event, with a column for my own thoughts and interpretations, allowing me to work to distinguish between the two.

Other documentation

This documentation is more difficult to quantify. It is also not as clearly visible in the emerging themes of the project. However, the information collected through these documents served to add 'thickness' to the analysis, as they served to support or deny themes arising from other sources. I made use of formal records of meetings in connection to teaching and student education, for example Student–Staff Forum minutes, weekly staff meeting minutes as well as less formal documents such as personal emails. Under this category, I also include public documentation such as website pages; student facing Virtual Learning Environment (VLE) documents as well as student assessed writing, feedback and other learning and teaching materials

Data Analysis

The data collection process was not structured in phases and relied on having open access to the Schools and participants. This access was less available in one (the STEM) site than in the other two, so analysis and the conclusions drawn here need to be viewed more cautiously.

The initial and main focus for analysis was the transcripts and notes from interviews and focus groups. It is hoped that the depth of information collected overall counterbalances, to some extent, the potential for a skewing of the data provided by participants who may or may not have had their own agenda in volunteering to participate. There were a few moments during interview when a subject requested what they said to be 'off the record'; this suggests that there was a genuine openness and willingness to provide honest responses to the questions posed throughout, but with a clear sense of the audience the research was likely to reach.

A good deal of reflexivity is required on the part of the investigator in order to find themes and develop an understanding of what the data collected can reveal about how those involved in teaching and learning understand the role language plays in their lives without asserting my own pre-conceived or previous-experience-based assumptions. The framework used for analysis was emergent, building on the usual models for ethnographic and case study research of allowing themes to emerge from the data through repeated reading, scrutiny and annotation. I initially attempted to use a template analysis (King, 2004). As the focus of the project was on teaching and learning, and also EAP I created a template that attempted to overlap the HEA's UKPSF (2011) with the BALEAP TEAP Competencies (2014), both of which outline professional competencies for teaching in HE, including required values, knowledge and areas of activity. This template clearly would not work for data provided by students, so for this I looked at the 'graduate attributes' as outlined by my institution and attempted to map threads onto this. However, working with multiple templates prevented any consideration of where teachers and learners intersected and shared experiences and I found I was trying to force threads into the template where they didn't really fit. This process was not wasted time though as the notes I made highlighted other more visible themes arising inductively from the data in the manner similar to that suggested by grounded theorists (Glaser & Strauss, 1967). I was strongly influenced by Academic Literacies as a theoretical framework (Lea & Street, 1998; Street, 1995) as threads that focused on socialisation, induction and resistance emerged. While the focus for analysis remained connected to language use, it became impossible to ignore data that gathered together quite strongly around other more socio-political, cultural and institutional issues relating to teaching and learning at TPG level. It is here that I found I needed to move beyond an EAP knowledge base and explore wider sociological concepts. Thus, different thematic layers become visible. Crucially, the dynamic interplay of the different elements of the HE context emerged as having a human impact on individuals in terms of their identity, sense of agency, sense of trust in their institutions and its structures and processes, and on their concept of time and time availability. Here again, I believe there is strong resonance with the experiences of others in other institutions, not only in the United Kingdom but across the globe.

Contexts

The institution

The research that forms the basis of this book took place in one UK Higher Education institution. The institution is a Russell Group University, so is research intensive and traditionally has been heavily research focused. It also positions itself as striving for excellence in student education, and as placing equal value on this as on research. Mid-way through the data collection period of this project, the University received a 'Gold' award in the first iteration of the TEF in 2017.

The University is large and diverse, with around 38,000 students studying across five different Faculties. More than 7000 of these students are classed as international. The majority of these international students study at TPG level.

The three specific sites chosen for the case studies were chosen to reflect some of the diversity that exists in provision, approach and student population across the institution.

The site of Case Study 1

Site one was the EAP teaching unit. This is a teaching unit; the academic members of staff are on teaching and scholarship contracts, with no remit around research and no requirement to submit to the REF (Research Excellence Framework). It is, however, structurally a part of a larger academic School rather than being housed in a service unit, thus teaching staff are on 'academic' pathway contracts rather than academic related or professional/managerial.

In comparison to similar units in other HEIs, this EAP unit is large, employing around 60 full-time teachers year round and recruiting between 60 and 100 extra teachers for the summer pre-sessional teaching period. Teaching takes place over the whole year; for most this is divided into four 10-week terms although some teaching is also semester based. Most (but not all) of the teaching undertaken by the unit is either on pre-sessional programmes that run throughout the academic year as well as in the summer, or on insessional programmes, where students already on their programme of academic study are provided with extra language development classes. All of the students taught by staff from this unit are from outside the United Kingdom and have English as an additional language (EAL)[5]. The summer is the busiest period of the year; in the summer of 2016, 2017, 2018 and 2019 the unit taught 1030, 1188, 1959 and 2500 students respectively.

During the early period of data collection for this project, the unit was developing and then delivering a new suite of summer pre-sessional programmes that moved teaching away from English for General Academic Purposes (EGAP) where students studied together regardless of disciplinary interest towards a range of English for Specific Academic Purposes (ESAP), where different programmes were developed in collaboration with academic Schools in order to better prepare students for their future disciplinary language and discourse. Most of the student participant volunteers were recruited from two of these discipline-focused pre-sessionals.

The site of Case Study 2

Site two was the AHC School. As with most Schools in this kind of institution, academics who work in this School are on a range of contracts, some who are focused only on research, some with the requirement to conduct research as well as teach, others on teaching and scholarship contracts.

For the purposes of this study, it is important to note that at the time of the investigation the School ran eight separate TPG programmes which had grown rapidly in size over the previous four to five years, largely due to an increase in the number of international student enrolments. In 2016/17 there were 332 students enrolled on its 8 TPG programmes; 46 of these were from the United Kingdom; 15 from the European Union and 271 International, with 180 from mainland China, Taiwan (3) or Hong Kong (2). Thus, it is clear that the vast majority of students studying in this School at TPG were not only 'international' but more specifically were Chinese.

Prior to the study, this School had worked to develop close links with the EAP teaching unit. It had identified language 'proficiency' as an issue amongst their cohort of international students and had created clear routes for language development support through the EAP unit. In fact, one of the new summer pre-sessional programmes was designed specifically for students entering this School.

Analysis of the School's external-facing website suggests high levels of focus on student education. Text describes the various programmes on offer and celebrates the successes of past students, highlighting employability as a key focus of the School's curriculum. When research is mentioned, it is in tandem with, and fore fronted by teaching: 'with teaching and research strengths …'. The website also highlights both University and School world rankings – something that is key to many international students' decision-making processes when choosing their place of study.

In summary then, this School was clearly focused on student education, and invested both time and resources in supporting this.

The School also had an international outlook; it worked to recruit and relied heavily on international students to populate its TPG programmes yet had also raised concerns about the language needs of these students prior to the study taking place.

The site of Case Study 3

Site three was also an academic School, this time based in a STEM Faculty. Again, the academics working within this School are on a range of contracts, with greater and lesser emphasis on teaching or research.

At the time of the study, there were 5 TPG programmes within this School/Faculty. Although each had its own clear identity, there appeared to be a large amount of cross over in module choice between the different programmes, so students worked together across the programmes as well as within. In the academic year 2016/17, 42 students studied at TPG level in the School. Of these, 24 were from the United Kingdom (with 3 from Northern or the Republic of Ireland; 4 from the European Union (excluding the Republic of Ireland), 14 were international from 11 different countries. In comparison to Site 2, then, in this School, at TPG level, the cohort was far more mixed, with around 1/3 of the students being international, and no language group other than English being dominant amongst the cohort. Its international student recruitment was not as extensive or as homogenous as that of Case Study Site 2, all of which may explain why no concerns had been raised about the language needs of its cohort in general.

Unlike Case Study 2, this School did not have strong links with the EAP teaching unit. Although a small number of its prospective students did take a pre-sessional programme, there was no clear focus within the EAP programming on the disciplinary needs of students entering this School and no formal collaboration between the two units.

The external-facing website for this School was, in contrast to Site 2, heavily focused on research. A form of the word 'research' was used 17 times on the home page, and the only mention of education was closely linked to research: 'our research drives the educational programmes we deliver ...'. This suggests a School with a more traditional Russell Group University focus, where research is key and viewed as the main driver for excellence. The suggestion is that students would be attracted to study in the school because of its research record rather than a focus on teaching and learning.

Participants

I have chosen to de-personalise the participants and use a code rather than pseudonym for all but two. This removes identifiable features such as gender or ethnicity that name choices can suggest, and that can in themselves lead readers to make assumptions around individual reasons for the comments being reported. When I feel it is important for understanding, I do identify a student as 'home' (to indicate they were from the United Kingdom and had English as their dominant, if not only, language) or 'international'/EAL. Whilst these sites of learning are clearly made up of the individuals who work and study within them, this manuscript aims to understand where language and content knowledge intersect within different sites of learning, and not to highlight the individuals themselves. The individual experiences and attitudes reported combine within one site to create the culture of learning that exists there.

However, at the same time, I also believe that something can be learned from the narrative of an individual, and that the experience of one student can provide important lessons. I have therefore chosen to highlight two particular students – Mai and Lin – one from each of the two disciplinary sites. I reconstruct a narrative around their individual experiences and use these narratives as a thread throughout the rest of the text to highlight key themes. In this way, I hope to demonstrate how an institution and the individuals who make up the institution can interact and have impact one on the other.

Who are the students?

In Chapter 1, I problematised and defined the term 'international' student for the purposes of this text. The understandings I am aiming to reach here around language and content knowledge relate most obviously and directly to these students, who are all studying through English as an additional language. These are also the students with whom I am directly concerned in my normal teaching practices. However, as part of the investigation, I also wanted to question this assumption. The Academic Literacies field of research is primarily concerned with students from non-traditional backgrounds who enter Higher Education through widening participation routes. Researchers in this field argue that the language of academic study is equally as occluded for these students as it is for EAL students (Ivanic, 1998; Lea & Street, 1998). I therefore wondered if and when language was perceived as a barrier by students in general as well as international students in particular.

I recruited student participants via a number of channels. The first channel was by visiting the classrooms of the cohort of students on the summer pre-sessional directly connected to Case Study site two. I did this early on in the teaching period and, through a verbal and written explanation of the project to each group of students, I received written consent from 155 out of a total cohort of 160 students to observe their classes and use their assessment pieces as part of my data collection.

Table 1 Coding for students

Site 1: EAP unit	Site 2: AHC school	Site 3: STEM school
Lin	Lin	
Focus Group 5: 5A; 5B; 5C; 5D	Focus Group 5: 5A; 5B; 5C; 5D	
Focus Group 6: 6W; 6X; 6Y; 6Z	Focus Group 6: 6W; 6X; 6Y; 6Z	
Focus Group 8: 8L; 8M; 8N; 8O; 8P	Focus Group 8: 8L; 8M; 8N; 8O; 8P	
	C1; C2	
Mai		Mai
		FS1; FS2; FS4; FS5

Towards the end of the teaching session, I then sent out an email to the whole cohort via their VLE asking for volunteers to meet me, either as a focus group or as an individual, explaining that I hoped to meet with them throughout the year that they were on their TPG programme. In response to this request, three groups of students and one individual (Lin) volunteered to take part in the project.

I recruited two further individual student participants from this site via a verbal and minuted request at a Student–Staff forum I attended and after a conversation at the end of an on-site seminar observation I undertook.

The student participants from site 3 were not as accessible. In this site, my attention was focused far more specifically on the experience of one student (Mai). Access was gained after a request for help with language was sent to the EAP teaching unit via her personal tutor. Other students volunteered to be interviewed individually in response to an email request for participants which was sent out by the Schools Student Education Support Officer.

Overall, therefore, the student participants from sites one and two were coded as in Table 1.

Here, I have detailed the individual participants who directly consented and provided interview and written data for the project. However, I would also like to acknowledge here the many other students who have indirectly contributed to my work, particularly those who have consented for me to observe and use their writing. Although I do not directly refer to their work here, I did consider it and it added depth to my understanding of the experience of developing post-graduate academic content knowledge through an additional language.

Who are the teachers?

As with the students, this project is concerned with the work of *all* University staff. Again, the concern here is the label used to describe us. There are many ways of distinguishing between the various roles we occupy, many of which denote some form of hierarchical status or leadership role. Many of these titles are also highly culturally bound, not only within UK HE, but more specifically within a particular University and do not immediately or directly translate to other institutions. This is particularly pertinent within roles which are centred on student education. For example, within the field of EAP, there has been some discussion (BALEAP JISCmail.ac.uk 03/2017) around the names used to identify with the profession, from Language Teacher, Lecturer in Applied Linguistics to EAP Practitioner, with many expressing a desire to distance themselves from 'lecturers' (because, naively, we do not 'lecture') and a tendency to label ourselves as 'other' than academics within a discipline.

However, although the names we choose for ourselves speak clearly about our own identity and position within the academy, most of these titles are also meaningless to, and possibly ignored by, our students[6]. What is important to our students in the context of this project is our role as *teacher* and how we contribute to their education. This is true regardless of the position we hold. Therefore, throughout this book, I will use the term 'teacher' to denote all involved in the education of students, regardless of status in the academic hierarchy. I will distinguish between the type of teaching undertaken by separating 'content teaching' (of a particular discipline or field) from 'EAP teaching[7], but when this separation is not relevant, the blanket label of 'teacher' will be used to encompass the interactions that any staff member might have with a student that involves learning taking place – including learning development and student support around, for example, careers advice and personal tutoring.

Within Case Study site one, the EAP teaching participants were all involved in teaching those students who were intending to, or already were, studying in Case Study site two. This included those responsible for the development and leadership of the summer pre-sessional programme as well as twelve of the thirteen teachers who taught on it; I met 11 as a group (LC3 – LC11) and another individually (LC1). I also met the six teachers who taught students in this School on the insessional programme as a focus group (LC1; LC2; LC9; LC12 and LC13), speaking to the programme leader both within the group and individually.

Via an email request, I received written consent to interview each of the programme leaders of the eight TPG programmes in site two individually as well as members of the School leadership team. One of these interviews took place via email, the rest being face to face and recorded. I also observed classes given by three of these teachers. These participants were coded from M1 to M11.

In site three, I observed three different teachers, and was able to interview one of these, as well as another senior member of staff. I also took part in meetings that involved one of the programme leaders and student support staff. I chose to code these as S1 through to S5.

As with the student participants, although the bulk of the data used for analysis was provided by the individuals listed here, the field notes I took throughout the investigation included commentary and observations based on interactions with a wider range of staff across the three sites as well as the wider institution. The intention of the project was not to highlight the experiences, approach and attitudes of individuals but to aim to create a nuanced picture of the 'norm circles' developed within each Case Study, where a norm circle 'consists of a group of people who are committed to endorsing and enforcing a particular norm' and they have 'the causal power to produce a tendency in individuals to follow standardised practices' (Elder-Vass, 2012 in Ding, 2016: 12). This is only possible to do by drawing from a range of data sources.

Chapter Summary and Practical Lessons Learned for SoTL

In this chapter I have provided an overview of the external influences on the development of this book, ones that are influencing Higher Education globally. This includes the movement towards evidence informed teaching and learning in Higher Education through scholarship. I considered the ethical implications of engaging in any SoTL project, and how this might impact both positively and negatively on both student and teacher participants as well as the wider institution. I also connected SoTL to the metrics involved in measuring teaching excellence and suggested that this can act as a push for institutions to support those who wish to engage with scholarship more formally, but also highlighted that with this push comes the need for specific support around developing expertise in SoTL methods and processes – giving my own case as an example.

The second half of the chapter then provided details of the scholarship project itself, positioning the researcher within the project, outlining the methodology, data collection and analysis methods as well as providing descriptions of the different sites of study and the participants involved. I end with key points of learning that can be taken from this chapter, and questions that readers, particularly EAP practitioners, might wish to consider for their own scholarship projects.

- You are investigating your own practice, much of which is likely to have developed through experience. You are therefore likely to experience a personal resistance to theory and theorising around your practice. My own approach was to be a magpie; to use a range of theories to explain different phenomena. It is necessary to engage with theory, but it is possible to begin with practice and look outward rather than to fix on a theoretical framework from the outset.

- SoTL itself is still under-defined and unclear. Use this to your advantage. Be part of the definition.
- Work on questions that are relevant to you, for the benefit of your context and your students.
- Collaborate. Draw on the expertise of others. If you have identified an issue that resonates with you, you will be amazed at the buy in you get from others.
- EAP teachers (and others) who often have a lower status within HEIs can, through SoTL and collaborations, make it clear that you have something to offer that others are not able to provide. It is, as Ding and Bruce (2017) have already argued, through SoTL that you are able to find your own academic authority and become a central part of a university's endeavour to improve student education.
- SoTL is an investment and commitment, and therefore requires investment and commitment on a personal level (as well as preferably from an institution). It is personal and the commitment is therefore emotional as well as intellectual. This is in equal amounts draining and incredibly rewarding.

2 Tracing a Student Journey: The Stories of Mai and Lin

In telling the stories of two students, Mai and Lin, I hope to create 'red thread' narratives of two student participants that weave through the rest of the thematic data presented in following chapters. In presenting this data, I try to use the students' own words as much as possible whilst providing my own understanding and interpretation of the background and context that created their own particular circumstances. I have chosen, from the many students I spoke to, to highlight the stories of Mai and Lin for a number of reasons.

Firstly, there are clear external parallels between their journeys. Both students were female and from mainland China. Neither had travelled outside China prior to the commencement of their studies in the UK. Both had only recently completed their undergraduate degree in China but had done some volunteer work within their chosen fields – Mai as a teacher within her discipline and Lin producing marketing materials for her province's internal tourism campaign. Both students attended a pre-sessional in the EAP unit prior to beginning their TPG programme, needing to do so in order to meet the language requirement of their academic programme because they had not met the overall 6.5 in IELTS requirement stipulated by their School's admissions policy.

The second reason for choosing these two students are the differences between them, the most obvious one being that they were studying in different Schools and different disciplines. While I am not suggesting that these two individuals can represent the experience of all students, or even all International or Chinese students in these different sites, their stories do highlight some of the conflicting tensions and possibilities that surround approaches to teaching and learning in an increasingly diverse higher education landscape. I have attempted to represent both students as unique individuals while drawing on key elements of their stories that represent the tensions within the internationalised higher education system. Remaining conscious of the 'danger of a single story' (Adichie, 2009), I have attempted to consider the story of the students' journeys from multiple perspectives, questioning motives, reasons and outcomes. I am confident that readers will recognise and be able to draw parallels between these stories and the stories of students they have encountered in their own teaching contexts. I begin with Mai's story.

Mai

Mai's story highlights the complex tensions involved in one individual student's situation as they struggle, and ultimately fail, to meet the expected academic and communication standards of their taught post-graduate academic programme. Through her story I hope to show that this failure cannot be attributed to one incident, to one breakdown in support systems, to a language deficit, a culture or educational system in isolation, or even to luck. Rather, each student experiences, succeeds or fails as a result of a unique and nuanced combination of factors. We need to consider these factors holistically; by breaking them down and trying to fix one aspect that we identify as being broken, we inevitably shift a problem onto a new or different area.

Mai began studying at my institution in January 2016. She arrived with a relatively low IELTS score, 5.0 overall, and had chosen to study in the United Kingdom for a year in order to develop her academic English via a longer pre-sessional programme rather than stay in China and continue to take repeated IELTS tests. She had had an offer to join her chosen TPG programme in the previous academic year which she had chosen to defer, I assume because she had not met the language requirement within the offer. She studied in the EAP teaching unit over three 10-week terms. At the end of each term, she submitted assessments and took part in tests which would allow her to progress to the next level of pre-sessional study. These assessments were intended to provide formative feedback to students that would enable them to learn and apply this learning as they moved onto the following terms of study, but also to provide an indication of their current language proficiency. The criteria used for the language element of the assessment was carefully mapped against both the CEFR and IELTS as these are the current most widely accepted units of measurement for language proficiency in the UK. Criteria also focused on the extent to which a student was able to fulfil the literacy and communication requirements of a specific academic task.

Mai struggled at each assessment point and progressed to the next level of pre-sessional with clear and strong warnings that she would find the following term difficult and may want to reconsider her choices and options for future study. All of the EAP teachers who were involved in working with Mai expressed at some point their belief that she would struggle with her chosen TPG programme and that she did not currently have the academic attributes that would enable her to succeed. Despite this, when measured against the criteria which were primarily focused on general academic language proficiency, Mai continued to progress towards her final goal of joining her TPG programme.

Having studied on three terms of English for general academic purposes (EGAP), for the summer, Mai was placed on a pre-sessional programme that was focused specifically on STEM disciplines. This was the first year that this STEM pre-sessional had run and the programme leader had reported difficulties in creating a cohesive programme that covered the needs of a still wide-ranging set of TPG programmes, covering five different faculties. The majority of the students on this pre-sessional were moving onto engineering programmes or food science and nutrition. Therefore, this is where the majority of the content focus was aimed. Mai was the only student on the pre-sessional who was going onto her specific programme, with only two other students moving into her Faculty.

Again, Mai met the expected level of proficiency at the end of programme, scoring 56 overall and no less than 55 in any of the components, with the standard expected level being 55 for students who, like Mai, were required to demonstrate a language proficiency level of IELTS 6.5 or equivalent in all four skills. Mai was allowed to join her chosen TPG programme, with the suggestion that she might want to consider continuing to develop her EAP skills by joining a general insessional programme. At the time, this insessional programme ran in four-week blocks of two hours per week, with students being asked to select the area of EAP they felt they most needed to focus on, the choices being: Academic Writing; Academic Reading and Critical Thinking; Academic Language Development; Academic Lecture and Seminar Skills; Academic Speaking and Presentation Skills; Grammar for Academic Writing. However, there was no absolute requirement either that the School act on this suggestion, or that Mai enrol for any of the classes provided. Frequently, timetable clashes would prevent Mai or any other student from being able to attend their class of choice.

The School that Mai had joined had very few structural links or connections with the EAP teaching unit. However, over the summer I had begun to have conversations with the Faculty's Student Education Service Manager and an academic member of staff who had taken on responsibility for all international students within the Faculty. The Faculty had very recently begun to consider the best ways to support international students, and I was being asked for advice around the development of a Faculty run series of classes for undergraduate students that would provide them with extra support around academic expectations. As a result of this contact, I was sent an email in early November with the following message from Mai's programme leader and personal tutor:

- *I have just marked her X piece and it is completely incomprehensible. I think her English is at such a poor level she does not understand anything. She has been in two coursework surgeries with me where she has said nothing and written very little and only spoken when asked a direct question. I spoke to her afterwards and she finds it difficult to understand what we say. She had a copy of the paper we were working on and she had made notes in Chinese all over it. She did not understand the (written) instructions for the flow diagram so did not bring a draft to the second coursework surgery and then when she sent me her draft it was clear she didn't understand what she was being asked to do, despite having sat through the two CW surgeries where we discussed the figures and the other students presented their drafts which we discussed. I had a personal tutorial with her and advised her to get more support through the language centre but I don't think this will be enough. She came through the presessional English course so I don't know how her English can be so bad but I really don't think she is going to be able to cope. (S3)*

Analysis of the content of this email highlights multiple intersecting and contradictory threads that I will return to repeatedly throughout the rest of this book. Within this message, there are questions raised about the following in relation to all, but specifically EAL/international students:

- Is language proficiency and therefore ability to linguistically de-code, the knowledge, instructions and tasks students engage with the main issue?
- Is academic ability – i.e. the level of knowledge that is required at TPG study – regardless of language a key issue for students?
- Do students have the required foundational knowledge in a discipline upon which to build the new information they are expected to work with?
- Do teachers approach face to face sessions in a way that helps students to feel comfortable and voice their (lack of) understanding? Are teacher's expectations fair and reasonable?
- Do teachers speak to EAL students in a manner and using language that enables them to understand what is being said and respond with ease? How do they check this?
- Are the instructions that students are required to follow clear and easy to understand?
- Is there any questioning around students' choices to make notes in languages other than the one of instruction? Should this be seen as a problem at all? Does it necessarily demonstrate a lack of understanding?
- What role do EAP teachers play in the success or failure of students once they move beyond the pre-sessional programme? Do pre-sessionals inadequately prepare students for a TPG programme?
- Is EAP assessment of language proficiency incorrect?
- Does/should taking a pre-sessional automatically mean that a student is able to cope with the language load of any TPG programme?

The surprise at Mai's language level also suggests that there is little awareness of what IELTS 6.5 or equivalent means in reality or of what difference can be made to this on a pre-sessional programme. This is a problem faced in many HE contexts (Ginther & Elder, 2014 for parallels within the United States and Australian contexts) It is clear then that the difficulties that Mai was facing were far more complex than simply not having good enough language that should already have been fixed by some extra EAP focused classes.

As well as agreeing to meet both Mai and her teacher to establish where help and support could be provided, I looked again at the assessments and grades Mai had been awarded by the EAP unit, and asked a number of colleagues, some of whom were also IELTS examiners, to provide second opinions of the writing she had produced. There was general consensus that the assessment she had received from the pre-sessional programmes was accurate. Comments on her writing were as follows:

- *The student has issues with linking ideas. She frequently uses commas to link ideas when it isn't appropriate and has issues with basic linking of ideas ... It is littered with basic grammatical mistakes ... I'm not sure these made the piece of work incomprehensible. (LC15)*

This second statement, from an EAP teacher, focuses fully on the language of the piece of student writing. Problematically here, though, the language is taken out of the context of the discourse being built. The EAP teacher is confident in commenting on basic grammatical errors but lacks confidence in connecting these to the comprehensibility or not of the piece of writing. The comment suggests, as might be expected, a strong focus on the mechanics of language used to create a low-level coherent utterance – language that is used to connect ideas within and across sentences to build a paragraph. What is missing is any suggestion that these words work to demonstrate any clear disciplinary knowledge and understanding. In fact, the EAP teacher finishes with this very comment – that she is unclear as to whether it makes sense here or not. There is then, a disconnect between what is valued in terms of language learning and accuracy and the clear demonstration of how to communicate disciplinary knowledge.

In both of these statements, therefore, language is viewed as separate from content but from opposite perspectives – the first viewing language deficiency as the cause of written incomprehensibility, the second viewing it as problematic but not necessarily creating incomprehension – the suggestion being that this may be as a result of lack of content understanding. Mai herself needed to move between and navigate through these two different approaches to the work she produced and the way she communicated.

- *I think in the Language Centre I learned something about reading skills, writing skills, something like that but actually I think in my post-graduate study requires more your academic knowledge.*
- *The language that I need is about my project, yeah. So maybe I know in the Language Centre the final, the, yes, the final language course is about your subject but I think I need more specific like how some study plan to let you know how to study the academic word, yeah. Maybe also give some website about your course maybe.*

In fact, Mai's School did provide online information to all its TPG students over the summer to help them to prepare for their programme

- *over the summer before the course started we were given some guidelines and we were given some suggested reading and also there was an online quiz that you could take multiple times and that was really helpful because that explained basically the background that they wanted you to be familiar with. (FS4)*

Mai should have received this, but when asked about it had no memory of receiving the information and had definitely not looked at it. While it is possible that she did not receive the information, what seems more likely is that she did not register its importance and was already fully focused on developing her language on the pre-sessional programme. This is what she had been informed should be her priority before beginning her TPG studies. By not engaging with this preparatory work, Mai was already less well informed than others from day one of her academic programme.

This bi-directional pull between developing content knowledge and language learning is a continued theme in Mai's story. Following on from the email of concern from Mai's tutor, a support package was put in place for her. This involved her being allocated a Post-graduate researcher 'buddy' who was part of an 'accelerated learning scheme' to help her review subject content, as well as being told to attend extra language development classes and 1:1 meetings with me to discuss her work. This was decided in a relatively formal, and minuted meeting:

- ***'It was agreed that [Mai] would rewrite her XXXX3M literature review and submit on Monday 21st November. This would not be reassessed but [Mai] would be able to work with MA and BB to write at a higher standard.*** *This will be a useful exercise and allow [Mai] to learn how to write an assessment at the expected standard. BB will also be able to help [Mai] learn associate skills and strategies to help her write future assessments. It was agreed that BB, MA and [Mai] would meet together in BB's office on 7th November. MA will meet with [Mai] regularly (ideally twice a week, for the next 2–3 weeks).*

As previously discussed, [Mai] should continue to attend teaching sessions but would be allowed to miss recorded lectures to attend Language Centre sessions, and other meetings with MA and/or BB. She should use lecture capture to catch up on missed lectures. She will also receive an extension for XXX2M data analysis I (15th December), which will give her two weeks following submission of XXX2M data analysis to focus on the XXX assessment.

Mai was, then, given extra time and support to allow her to develop her understanding of the necessary skills and content, but was also told to prioritise language classes (in isolation to her subject content) over lectures that she could catch up on later by watching recordings. In fact, the message that Mai took from this meeting was that she should attend these language classes as a priority over all other sessions. She then went on to miss a number of sessions that were not recorded because they clashed with the language insessional classes, despite her conclusion that:

- *actually I think the language course just help you with the writing skills is very useful and other just to learn, like let you know how to live in UK ,yeah. Some, I think just some life, some were life, some skills about how to live.*

Furthermore, in conversation with me, Mai was able to identify that what she actually felt she needed help with was technical, subject specific vocabulary development. Mai tried to express how this was one of her greatest barriers to learning in multiple ways:

- *maybe sometimes I can't understand the words. Okay, I am still thinking what they mean about the word. And what the teacher said later, I don't know, yeah. I maybe not care.*
- *The most difficult I think is the academic word, yeah, because in some about science project maybe have some more academic word, yeah, you need know and you don't understand … academic biology word.*
- *when you reading the word you need record it and know the translation and then if someone like teacher maybe said this word so you know the word mean.*

However, Mai barely engaged in the support she was offered that did move towards connecting language work with her academic communication needs. Whilst she was happy to meet me to chat and be recorded talking about how she felt about her studies and language development, the academic work that we were supposed to discuss together never materialised. She had always forgotten to bring it or had not managed to write the paragraph we had asked her to re-work. We did have one meeting that included Mai, myself and the 'buddy' (MA) but follow up meeting requests went unanswered.

So, in spite of being offered the support she identified as needing, and had been told to access by her programme leader, Mai chose to take up the more surface level opportunities for engagement rather than ones that might have enabled her to extend and deepen her understanding.

There are examples of this surface level engagement running throughout the data I collected around Mai, covering a range of different aspects of her academic life. From my own notes of the first meeting I had with Mai and her tutor:

- *For the assessment, Mai had generally completed the parts of the paper that required her to follow instructions, copying & pasting sequences from a website. She had not attempted the higher order tasks requiring critical analysis or thought.*

From a further chat over coffee that I had with her, I recorded my surprise at her lack of personal connection with people in her School

- *She cannot remember the names of any of her tutors! She is getting a new personal tutor – she feels that XXX was always too busy to speak to and support her. She asks her for help and she says she doesn't have the time.*

It is here that many of Mai's problems seem to lie. Her tutor had, very early on in Semester 1, identified Mai's need for help and acted quickly to try and get something in place for her. However, this support did not fit properly with Mai's specific needs or with her timetable and did not seem to meet her expectations of the kind of support she wanted nor from whom it should come. Beyond these actions, the teacher did not make herself available to provide extra help as she did not see this as part of her responsibilities. The tensions between student need and expectations and academic time and expectations, and the tensions between language proficiency and academic content competence all come into play. Mai, it seems, fell between the gaps of each of these and was able to suggest many reasons as to why she was not managing to complete the required work, or understand.

It was, in part, connected to the intensity and density of the content, delivered in an as yet unfamiliar language:

- *if I listen the English a long time maybe I have some headache, yeah. So maybe in one hour of class I just listen half an hour.*

It was also a lack of subject specific vocabulary:

- *The most difficult I think is the academic word, yeah, because in some about science project maybe have some more academic word, yeah, you need know and you don't understand ... academic biology word.*

Yet also connected to perceived differences in levels of knowledge, and assumptions made around the knowledge she had:

- *I think my classmates have knowledge much more than me ... I think because they are UK student, you know. They, maybe they know how to study in UK but I don't know and maybe their knowledge is, some knowledge that we learn is different. It's some may be a little different, yeah, and maybe the, and the teacher know what they learned before but the teacher don't know I learned before.*

Here, Mai also touches on the differences in educational approaches to study, and what knowledge is privileged, and this leads to suggestions that confidence is also an issue. Here Mai initially suggests that confidence is an issue for all international students, but quite quickly moves towards it being a personal difficulty:

- *the international student don't have confidence to join the class, to join the, to join to talk with our UK classmate.*

  BB: *Why do you think that is?*

  Maybe just my problem because I am very shy in the class so I have some question maybe I just ask the classmate who nearby, yeah, to me. So, and use a very low voice to ask, yeah, because I think maybe they all know this answer but I don't know, yeah. Yeah, it's just about confidence.

However, behind all of these self-identified barriers to being able to participate in and gain access to the learning community she is part of, there is also a resistance to doing so. This seems to be particularly true around developing an understanding of the specialist English that she needs in order to do this. Mai reports this as 'not caring', but it also seems to be a (semi) conscious resistance to the use of a different language to communicate with peers when Chinese is viewed as the dominant and common language for those she will remain in contact with. Therefore, when discussing English as being the accepted language for scientific communication, Mai's response was:

- *Maybe in the, maybe in China they give the Chinese name and after the Chinese name give the English name, but most of Chinese student...*

  BB: *Don't listen?*

  ...they, yeah. We are missing the English name. Who care? I think, because they all use Chinese. They don't know the English name.

Despite having been given the opportunity to learn this key, core technical English vocabulary prior to beginning her TPG studies, Mai had resisted learning it and continued to use only the Chinese. Mai also reported choosing to discuss her studies in the United Kingdom only with other Chinese students, including a friend who was studying in China, rather than those from other countries on her programme. As there was only one other Chinese student taking the same programme, this limited her choices considerably:

- *my classmate from China so sometimes maybe we can together, sometimes because some option class is different, maybe do it myself.*

  ***BB:** And what about all the other students? Why does it need to be the one from China?*

  Because we can discuss more fast yeah. If I don't know how to say it use English I can use other word, you know. And if with other UK student, sometimes we make maybe, the speed is so fast.

In fact, Mai expressed as much concern over not knowing the Chinese term for something as she did over her lack of English vocabulary:

- *sometimes if you use English you can't remember the Chinese name.*

Her concern over maintaining her identity as a speaker of Chinese meant that she maintained a focus on translation as a means of understanding the knowledge that was communicated to her in English. Thus, when reading, she continued to translate all unknown vocabulary into Chinese, keeping a notebook that was simply a list of words with their translation:

- *when you reading the word you need record it and know the translation.*

This meant that in class, she tended to not pay attention and to simply wait until it was finished to translate or ask her friends to explain, ultimately leading her to suggest that she stopped listening and, again, 'didn't care':

- *It's difficult. I need time to understand so I don't make mistakes. I will do after class. I remember last class like this, it takes a lot of time. I will ask my friends and they explain.*
- *maybe sometimes I can't understand the words. Okay, I am still thinking what they mean about the word. And what the teacher said later, I don't know. I maybe not care.*

This resistance to full engagement with developing the language needed to communicate effectively in English in her discipline could be seen as an agential choice, a desire to maintain a previously held identity and a push against a powerful hegemonic force. It could also be seen as a way of protecting herself, by saying she didn't care she was able to save face when she didn't meet the requirements of her assessments. Either way, Mai presented a challenge to this School. It was clear that the majority of participants from the School had thought very little about the impact and use of language in communication of disciplinary knowledge. One of the teachers I observed, for example, argued that language in his context was not an issue at all; there are just a few key words and that from these enough could be understood. Yet during my observations I wrote down lists of conceptual vocabulary and language used in a subject specific context that would be likely to form a barrier for an EAL speaker who had previously studied English for general (academic) purposes. During the practical classes, there was a great deal of general conversation, both around social activities and the academic work being undertaken. The two were frequently interwoven, making it difficult to separate where one ends and the other begins. For a student struggling with both the new content of the subject and the language of study, the obvious approach to this 'noise' would be to ignore it, to block it out and to decide you 'didn't care'. I observed Mai do this and by doing this much of the expected or presumed learning was lost, as was much of the affirming and confidence building that is important for avoiding a sense of isolation when struggling with information.

Although Mai herself seemed to resist incorporating language development into her disciplinary studies, this lack of awareness of the importance of language from those who were teaching her only compounded her difficulties. Mai, overwhelmed by the amount of work she was presented with, relied on her teachers to guide her as to what was important to learn. By either ignoring language as an issue at all, or suggesting it was something to learn elsewhere and not an integral part of her subject, Mai was not given the right opportunities she needed to enable her to succeed. A shift in this perception is clear in a later email I received from Mai's tutor:

- *Any chance you are free on Wednesday morning? This might be an opportunity to get to the bottom of whether it is primarily a language problem- [Mai] not being able to understand what she was being asked to do – or a combination of language plus science – doesn't understand the concepts so can't articulate them.*

However, by this time it was too late. Mai's resistance had become even stronger and, although she remained determined to continue with her studies, she no longer responded to offers of support from others. She was given extensions to assessments but did not achieve the grades that would allow her to successfully continue into Semester 2.

Lin

In contrast to Mai, Lin's story is one of success. However, as with Mai's difficulties, this success cannot easily be attributed to one specific area or personal attribute. Rather it was a combination of agential choice, strong identity, resistance, support and structures that combined to enable Lin to navigate through her programme with success.

Lin, like Mai, began her time at University in the United Kingdom on a pre-sessional programme in an EAP teaching unit. She arrived with an overall IELTS score of 6.0 and took the summer content-based pre-sessional in order to make up the 0.5 shortfall in her language entry requirement. While Mai's pre-sessional programmes had either been English for general academic purposes or, over the summer had moved towards having a science focus but in a broad sense, Lin's programme had been created in collaboration with a teacher from her receiving academic School. The content of the programme focused on foundational theories and knowledge for her discipline and the assessments she took were written to enable her to develop an understanding of her disciplinary discourse norms. At the end of the pre-sessional programme, Lin achieved an overall score of 60, which was 5 marks above the agreed expected level for entry to her TPG programme.

On the final day of the pre-sessional programme, the EAP unit held a 'transition' event. Through this, students presented their developing understanding of their own identity as students of their discipline, relating this to key theories of identity that they had studied on the programme (see Bond, 2019). Teachers from their receiving School attended this event, and also took part in a symposium where they answered questions from the student audience. In this way, a clear link between the teaching and learning on the EAP pre-sessional programme and their future TPG programme was established. The language and discourse studied was centred on that of the discipline. This connection continued into the academic year as an EAP teacher was employed to work specifically in this School, to teach insessional EAP classes that supported the TPG programmes and offer individual consultations. Thus Lin, as well as all her peers in this School, had continued access to language support throughout the year, and the messages she received from both EAP and subject teachers were consistent around the importance of language use and communication in her discipline.

Within the School, in contrast to Mai's context, there was a consensus around the difficulties that language created and an understanding that the discipline was centred on language use and nuanced argument created by its expert manipulation. Whilst teachers in this School, as in Mai's, did not see language work as part of their own remit, they did see a need to collaborate with EAP tutors to enable students to develop a discipline-specific language knowledge.

While Mai found herself on a programme that had a small cohort (8 students) from a range of countries and backgrounds yet chose to develop close relationships only with the one other Chinese student on the programme, thus avoiding the need to communicate in English, Lin was on a programme with 53 students enrolled on it, 44 of whom were from mainland China. For Lin, the path of least resistance would have been to establish partnerships and relationships with other Chinese speakers. It was clear from the beginning of my contact with her that she was determined not to do this.

Lin contacted me after I sent an email out to all the students who had completed her pre-sessional programme asking if they would share their group presentation from the final, transition day with me and possibly agree to come and speak to be about their transition onto their TPG programme. Other students who responded to this email request did so as a group and came to see me as a focus group participant. Lin emailed me as an individual and requested to come and speak to me as an individual. She told me that this was because she felt that she had done all the work for the presentation herself, and that her group members had either not bothered to attend the presentation or had done so but not really participated. She expressed both anger and disappointment at this behaviour; she felt it was rude and wasteful but declared that many Chinese students had come onto the programme with that attitude. Lin frequently described the behaviour of other Chinese students in negative ways as a means of separating herself from them and of creating a space and identity for herself as a serious student. The most extreme way of doing this was to suggest that some of her peers were involved in contract cheating:

- *I should add that some people who travel here they chose the same person and usually this person will try to keep their score in low level, so they try to avoid those lectures, focus on them … so they will not compete with students who want to do well so we don't have complications, therefore we usually, we're not talk about it or … Because they don't have conflicts with other people so we just ignore them.*

While Mai, largely isolated from others who spoke her home language, resisted adopting and developing the language she was studying in, Lin resisted being seen as part of a homogenous 'Chinese' student population. When I observed her TPG classes, Lin had clearly made a conscious effort to establish contact with the non-Chinese members of her programme. In both the lectures and the seminars that I attended students were asked to work in small groups. The majority of these were by necessity all-Chinese groups but Lin worked to ensure that she was in a group that contained a wider range of nationalities.

This is not to say that she isolated herself entirely from other Chinese students; in one interview she described how they supported each other to participate in seminar discussions by code-switching and translating key vocabulary for each other so that they didn't become stuck when:

- *you have mass of things in your mind, I know this, I know that and you just don't know how to express it out, because sometimes you need an English word but you only have a Chinese word and sometimes they reverse.*

Lin also describes using translation as a tool for understanding. However, her approach to this was different to Mai's. Whereas Mai used translation to shift English texts into Chinese to enable understanding, Lin's approach was to use translation both as a tool for learning, for note making and a way of avoiding plagiarism:

- *I translate them into Chinese and transfer the Chinese into English…*

Her approach more closely matched Canagarajah's definition of translanguaging as a 'naturally occurring phenomenon for multilingual students' (2011: 402), which can lead to codemeshing as the textual realisation of translanguaging practices. From these notes, Lin then began to develop her writing and also her vocabulary, describing a sense of frustration that she knew from her reading that there was a 'better' word for the choice she had made in her translation back into English, but not feeling she was yet proficient enough to use it correctly. However, she was also able to track her progress in understanding and see the connection between her developing understanding of her subject and of the language used to communicate it.

- **12/10/16** *I'd like to share my study process as well. Firstly, I check my handbook to figure out what the topic is in this week and write down some questions I have when facing this topic. Then, I use a memo to record how many pages I need to read and I have read. This helps me less stressful about the reading task. During the reading, I use the dictionary to record how many fresh words I've met I'm always happy that I find I have less unknown words than the last reading. I also list my questions in the reading. The lectures usually help me answer those questions. If I still feel confused, I will figure out them in the seminar.*
- **27/10/16** *I changed my reading methods this week. I focus more on finding what the argument is in the article instead of trying to understand every word. I used to spend three hours to read one article but now I only spend one hour. How many times I spend on an article depends on the topic. If it is highly relevant to my essay, I will read it more times. Usually, I get the main idea for the first time reading and then I read it for details. Sometimes, I may read it again while comparing with another article.*
- **7/11/16** *I'm confident to find the arguments in articles. I think it's very useful to focus on the hint word, such as 'I agree that…', 'I find that'.*

- **9/12/16** *This week, I went to see [my tutor] and had a talk about my essay feedback. I realize my ability to seek resources should be improved. For example, I didn't use any journal article in my essay because I didn't know how to find them in the library. All I read are books and It is really low efficiency. But now I know how to find articles accurately by using advance search. I have to say the understanding of language sometimes have a negative influence when I try to search some materials. Because I do not know what the key word is in English. For example, I spend a long time to find the word Comic-con which is going to be the key word in my dissertation. I know the Chinese word but it's really difficult to find the match word in English.*

Although Lin identified similar issues (conceptual and disciplinary vocabulary; ability to express thoughts and ideas in front of others; reading complex texts), her experience in facing these difficulties was different. Mai became increasingly isolated and disengaged, finding the University systems and staff incomprehensible and unsupportive of her needs. Her knowledge of her discipline and her ability to communicate in English, or at least her confidence in both, seemed to diminish. Lin, in contrast, found a way to focus on what she did understand and expressed a confidence that this would ultimately enable her to learn and achieve her goals. She felt a part of her learning community and was able to make use of people and systems in order to develop her understanding. While Mai felt that her tutor did not have time for her, Lin made use of her lecturers, but also saw the value of talking to other students and support staff and asked targeted questions having identified where she felt her own current difficulties were:

- *the Lecturers do give many useful advice and they will explain the … the target, why they ask us to do this essay and … I think the most important thing I learned is, how to search sources … I just don't know how to find the materials I need but now I know that I can use Google Scholars, how to use the advanced search in the library and now I'm more efficient to find those articles.*

Lin also recognised that she was learning concepts that she did not have the vocabulary for in her first language. While this became an issue for Mai who continued to try to separate language from content knowledge, Lin was more pragmatic and accepting:

- *Sometimes I will, separate them, because I don't know the theory just because I don't understand the language, but actually I have learned the theory before, but sometimes if I completely don't know the theory before, and … I can learn from the Lecturer, so basically it's like, the theory is a new word for me and I don't know the theory before, so actually I don't know the Chinese meaning of this theory but I know what it meant.*

The agential choices Lin made are evident even in relation to her communications with me. Initially, she chose to contact me via email, sending me a message each week outlining her thoughts and answering my subsequent questions. She chose this method because she wanted to practice writing in English; she explicitly shows evidence of learning through this communication here:

- *I hope you enjoy the weekend as well. (Actually I never know how to end an email politely, just learn from you.)*

After she had submitted her first written assignment, she decided she would prefer to speak to me face to face. This decision coincided with an assessed group presentation, but also the next phase in her trajectory through the academic year.

- *Since I have already submitted it, I think it is useless to struggle with it ☺. I read the assessment criteria before I started my essay. I think I have tried my best to achieve it. I guess my score might be 60 to 65. I'm quite confident about the breadth of my reading, and I do have some of my own arguments in the essay. And I also have a clear structure. But I'm worried about the evidences to support my arguments and the depth of my understanding towards some literatures. And I'm also worried about grammar mistakes and whether the vocabulary I used is academic or not.*

In this email, Lin shows a clear understanding of the study skills she felt she needed to be successful on her programme and she was accurate in her own assessment of the grade she would achieve on the assignment. She also showed a clear understanding of where she needed to go next with her learning. However, she was clear that this would develop as she continued to study, and now felt that she had the time, confidence and capacity to explore other, social, opportunities offered by the wider University:

- *At first I just, focussed on my studies because it was a big problem at the time, but later on I find that I just don't want to communicate with English speaking people because I think I have read too many English materials and I just want to stay with my Chinese friends, and after this period, I started to want to, again, to make new friends but most of the English people already have their communities so they don't accept a new entrant.*

So, although by January Lin felt ready to move beyond her formal study programme and begin to interact and have an impact on people and structures outside her taught modules, she was also concluding that it was probably too late for her to do so; that in an intense one-year programme, these kinds of relationships had already been formed and fixed within an academic year cycle that is founded on the premise that friendships, social relationships and participatory choices are made during induction events at the beginning of the year.

Chapter Summary

The stories of Mai and Lin are unique to them yet can also be seen to typify the conflicting stereotypes of the international, and more specifically Chinese, student studying, not only on a TPG programme in the United Kingdom, but in many HE contexts at any level of study. Their stories touch on a complex combination of support systems, expectations, language issues, reticence or fear of speaking up in formal learning environments, difficulties in understanding instructions and systems, isolation, assumptions around knowledge building, cultural privilege, determination, social interactions and (non-)integration. The differences between their trajectories cannot be pinpointed to one behaviour, attribute or external intervention and are not in fact specific to international students. Lin suggested a key factor in the different behaviours that teachers noticed in formal learning environments:

- *I think that because we have different object, some people come here like just for a vacation … And some people are here to learn something new … I think some people like me, they feel they have improved and they prefer to communicate with each other about their experience, but some people just stay with what they're used to.*

While there is little that can be done about individual student choices and attitudes towards the learning process, it is also clear that, even with students like Lin, there is more that HEIs could do to ensure they are fully included and have full access to all aspects of the curriculum. It is therefore necessary to consider all aspects where interventions and changes can be made to ensure inclusion. The stories of Mai and Lin suggest that this cuts across a number of areas and can be viewed from a number of perspectives that draw out interweaving themes. Therefore, in Chapters 3 to 7, I move beyond these individual student narratives to consider emerging themes from a wider data set that reveal the complexity of the interaction between language and disciplinary knowledge communication. From this, I suggest practical interventions that could be employed to enable teaching and learning practices in most HE contexts to become more inclusive through a purposeful focus on language.

3 The Taught Post-Graduate Curriculum

Having provided the narrative of two students' experiences over the period in which the data collection took place, I now build a picture around the wider context in which these students were studying. In this chapter I outline my understanding of the themes that emerged. Taking a critical realist perspective, I position the TPG curriculum as the socially real context within which all the actors interact and react. I then consider the impact these actions and the institution itself has on the ephemeral concepts of identity, agency, trust and time of the participants, with a focus on language being woven throughout. In doing this, I must also acknowledge that, having continued to work and practice within the same arena since the data collection process was completed, my thinking and the interpretation I place on some of this data and the context has in all probability adapted and changed as a result of my own interactions and actions since the end of the data collection period. I have re-cut and re-analysed the data and thought about it in a variety of ways. I have continued to interact with the curriculum beyond the period of data collection.

The taught post-graduate curriculum was the contextual space in which this study took place. It was the obvious place to focus the study because it is at this level of tertiary study where, currently, the majority of students are from outside the UK. To give my own institution as an example, census results show there were 4821 full time TPG students in total, 2995 of whom would be classed as 'International' in the academic year 2016/17; this figure contrasts sharply with the 1523 UK-based students. Nationally, 118,435, or 42%, of all TPG students studying in the United Kingdom came from outside the European Union with almost one third of these coming from China while 31,320 were from the European Union but non-UK domiciled in 2016–17 (UKCISA, 2018). According to HESA, in 2017/18 there were 458,490 non-UK students studying a first year of study across all levels of tertiary study in the United Kingdom, again by far the majority of these coming from China. Also worthy of note is the rapid growth in the number of students coming from China to study in the United Kingdom – rising from around 25,000 in 2006 to more than 75,000 in 2017/18. This suggests a need for rapid adjustments across the sector in terms of space and accommodation as well as approaches to knowledge exchange and language awareness.

With such large numbers of students coming from outside the UK, and therefore presumably from outside the cultural, social, linguistic and educational norms of the host institution, it would not be unreasonable to assume that the interaction between these students and the structures of the institution, its curriculum and the staff responsible for framing and controlling (Bernstein, 2000) the knowledge within the curriculum would lead to some kind of change or elaboration of the curriculum over time. To explain further, through the lens of critical realism (Archer, 1995; Archer *et al.*, 1998), the taught post graduate curriculum is 'real'; that is, it is the underlying mechanism that can only be explained through the events and experiences that occur as it is enacted. This is contextual and based in practice and, as Archer (1995) would argue, is either reproduced or elaborated over time as a result of the sociocultural interaction that takes place within its structures, i.e. morphostatis (reproduction) or morphogenesis (elaboration) occurs. At a TPG level of study in the United Kingdom, these socio-cultural interactions involve large numbers of students for whom English is an additional language and who are viewed as 'international'.

This chapter looks, therefore, at how the TPG curriculum is conceived and understood by those who learn and teach within its structure. It considers the impact that this highly international interaction with the curriculum has on individuals in terms of: their identity; their understanding and use of time; their trust in the curriculum; and in terms of their ability to make agential choices.

Firstly, however, it is necessary to define what I mean by 'curriculum'.

What is a Curriculum?

Approached from a language teaching perspective, there is a wealth of literature that defines the difference between the often-conflated concepts of 'curriculum', 'syllabus' and 'approach' (see, for example, Clarke, 1991; Kumaravadivelu, 1994; Meddings & Thornbury, 2009; Munby, 1978; Richards & Rodgers, 2001; Van Lier, 1996). Richards provides the following definition of a curriculum as: 'the overall plan or design for a course and how the content for a course is transformed into a blueprint for teaching and learning which enables the desired learning outcomes to be achieved' (Richards, 2013: 6).

He goes on to suggest that language teaching builds this curriculum in one of three ways, through either a forward, a central or a backwards process. The forwards process begins with the syllabus, considering what should go into each learning episode, including decisions involving corpus research, language acquisition theories and topic choice. This drives the assessment object that comes at the end. A central process focuses on a teaching method or approach, where the syllabus and outcomes are seen as resulting from the classroom interactions. The backwards process begins with the desired learning outcomes. Arguably this backward approach is the least common one taken, largely due to the prevalence of the more syllabus driven course books that control many language teaching contexts. The notable exception to this is when a book or syllabus focuses entirely on preparation for language tests such as the Cambridge Advanced or Proficiency tests.

Within EAP, clear parallels can be drawn with more general language teaching practices. Again, many EAP contexts remain coursebook driven, with the curriculum being built around the pre-selected content. Bruce (2011) devotes a whole chapter to syllabus design, and much attention has been given to materials development in EAP (Harwood, 2010) with decisions being made around linguistic theory, discourse practices and genre analysis. Although EAP also highlights the importance of being driven by student need and of enabling students to enter their area of academic study with the confidence, language and skills necessary to access their content curriculum, there are fewer published examples of backward process curricula in EAP. This may be because of the influence of IELTS as a marker of learning outcomes, so published backwards process syllabi are based around IELTS tests. I argue that this is not representative of EAP teaching. Another reason is possibly because a backwards curriculum in EAP is, by necessity, localised and contextualised within a specific institution and may be where EAP practice is more obviously influenced by HE practices rather than language teaching norms. A content-based EAP curriculum should probably, therefore, fit within the backwards curriculum, with learning outcomes being fixed around students being ready for their academic programme; however, if the 'topic' becomes the central concern, or the teacher who enacts the curriculum lacks the necessary knowledge to navigate the content and breakdown the discursive features and relies too heavily on specific language teaching methodologies or approaches, the local enactment of a curriculum can quickly shift it from one process to another.

In HE more broadly there has, however, been little attention paid to curriculum theory. Bovill and Woolmer (2018) suggest that this is due to confusion about what the term 'curriculum' means, despite it being the concept that underpins all university learning and teaching (Barnett & Coate, 2005) The seminal work of Biggs and Tang (2007) on constructive alignment has been the focus of learning design, but this has generally been at syllabus level, within modules. The modularisation of programmes has led to discrete item planning, with the intention being that students are able to build their own programme from individual modules. In this way, learning outcomes and assessment processes focus on developing discrete areas of knowledge and rely on students being able to make connections across and between their chosen modules. Recently, the focus has moved towards a more holistic view of student learning, and education developers are encouraging a broader focus on the curriculum (see, for example, Fung's 'Connected Curriculum', 2017a). Alongside this move towards planning a tertiary level curriculum, rather than a syllabus made up of discrete modules, there has also been a great deal of work around developing a curriculum that is inclusive, de-colonised and internationalised and that positions students as partners in the process rather than as passive receivers of proscribed canonical knowledge.

It is clear then that an understanding of the curriculum in higher education needs to be expanded beyond selected disciplinary knowledge to be learned, beyond the planned learning episodes, linked to final outcomes, beyond the *written* curriculum that is, largely, controlled by the academic teacher who developed it. For Barnett and Coate (2005: 51), curriculum 'is dynamic and in flux and is also a site of contested interpretations. A curriculum is fluid and is not – cannot – be caught in any schema or template'. The curriculum includes methods and approaches taken to the teaching, learning and assessment of the selected knowledge and the variety of opportunities provided to students to develop and apply this knowledge in different contexts. The curriculum should also take into account what has been termed the 'hidden curriculum' – the opportunities and barriers that are created in the social sphere of university life. This includes the support provided by, for example, the library, careers or counselling and wellbeing services, by the International Students Office and the opportunities for social interaction or isolation that come from Student Union activities or living in University residences. In other words, the TPG curriculum is seen to encompass all interactions and relationships that may have an impact on learning over which the University has some control or influence. Particularly when students are entering a context where there is an expectation on them to live and communicate in a language and culture that is not their own, the curriculum extends far beyond the classroom, beyond the planned outcomes of a written programme into the everyday and the unofficial learning spaces of the university. A curriculum, then, is seen as a collective, socially situated and socially real enterprise. A curriculum can be planned, organised and written but it also evolves based on prior iterations and the reactions and interactions of those involved. It becomes a social and cultural artefact (see Archer, 2000). When a curriculum is planned, it is done so with a specific group of learners in mind; when it evolves it is because of the ways in which these learners interact with the learning opportunities created by the curriculum as it plays out in reality.

Conceived in this way, the University itself is the curriculum and, for EAL students, is also effectively a language classroom, where language is viewed as the loci of struggle. This struggle takes place on multiple planes: in Foucauldian terms of power, in the sense described by bell hooks (1994) where (academic) (English) language works to both disrupt and oppress, and more literally as students grapple with the difficulties of language manipulation. More specifically the university curriculum can also be an EAP classroom, where EAP teachers can mediate for students between all of these learning spaces and where students need to act as ethnographers of the academy and need to find ways to navigate through a new context and culture. Students are both provided with and denied opportunities to develop and use their target language across a range of contexts. While students make their own choices as to whether and when they will communicate with others in this language, the University curriculum should ensure that this choice is there to be made.

However, this holistic view of a TPG curriculum steps outside the main concerns of academics who are focused on teaching and assessing students within a specific discipline, and on developing their students' understanding of a specified knowledge base. It is not, therefore, necessarily how the curriculum was conceived or understood by participants in the project. Here, key differences seem to stem from the interaction of two separate concerns. The first is the nature of the discipline being taught; this includes the epistemology and ontology of the field, but also the perceived status of the discipline within the academy. The second is the number of students enrolled on a programme, specifically in connection to the extent to which the student body was already 'internationalised', and the impact this had already been seen to have on teaching and learning practices.

What is the Purpose of a Taught Post-Graduate Programme? Who is it For?

Although the UK Quality Assurance Agency has clear cut suggestions as to the attributes that a Masters level student should be able to demonstrate, this clarity is 'not always fully demonstrated' among UK academics (Brown, 2015: 174), particularly in relation to the difference between undergraduate and taught post-graduate learning outcomes. This conclusion is supported by many of the comments from my project participants.

In part, this blurring is to take into account the need for all students to transition from one level of study into another, and an understanding that the knowledge base of many cannot be assumed to be that of the feeder undergraduate programme from the same institution.

Therefore, a teacher who was new to the University explained why STEM students who had taken their first degree in the same School were not required to attend his practical sessions:

- *the module that I teach the Masters students is very similar to a project that second year under graduate students, there are a few bits that we've added so in terms of complexity it's more Masters level this module but it's very similar so I think the decision was because maybe they might find that bit repetitive and so they would do something else. (S1)*

However, he was also clear that there were expectations for all students to have some practical and theoretical knowledge prior to commencing their Masters programme:

- *we assumed some of the techniques so we probably expect them to have done things like use a pipette and maybe done some calculations before but in the first week, week zero induction week, we got them to do a pipetting exercise and I hope that during our classes we encourage them to ask questions if any student is struggling both me and the demonstrators and the module manager. (S1)*

Within this discipline, then, a Masters degree was seen as an extension from undergraduate study, with extra depth and more skills being added:

- *We also expect the Masters to have some of these techniques already from their undergrads so they're a bit more independent working and then we've added a few extra techniques at the end and in terms of the assessment as well so that the lab report that we get a Masters student to do, the undergraduates don't have to do a lab report of the same scale, so I think it does go deeper for the Masters level. (S1)*

This is also how a student from the same School understood the transition, describing how teaching began as similar to undergraduate classes and then moving towards much more autonomous, research work in laboratories:

- *this first six-month block of it with the teaching is very similar to under graduate and then when we do, in April, we go into the labs for three months and that will be exactly like doing a PHD just for a much shorter period of time. (FS5)*

This expectation fits with the general understanding around STEM disciplines, where knowledge, rather than knowers (Maton, 2014; Bernstein, 2000) is the epistemological focus, and where that knowledge follows a seemingly linear path, adding, as S1 explains '*extra techniques … and going deeper*'. Within this site, it was clear both from the student-facing website, the approach to teaching and learning, the written curriculum and the students themselves, that a Masters degree was conceived as a linear path that allowed students to make the transition from Undergraduate study and then into a research degree. It did not really stand alone as a fully developed curriculum; rather it slotted into a four to seven-year programme, developing skills that built on previous knowledge and moved towards the next.

This fits within the traditional understanding of a University education (Barnett, 2015), with a focus on preparing traditional students to become a traditional research-active academics. However, it does not take into account the different knowledge bases that students may come to study from. As Mai experienced.

Mai found it difficult to connect her previous knowledge base to the content of her Masters programme, and it was clear that without the expected foundations, she was finding it increasingly difficult to access and interact with any of the knowledge she was encountering. This initially suggests that consideration needs to be given, either to the entry requirements for the programme, with a clearer understanding of how undergraduate qualifications transfer between countries, or that the curriculum needs to be broadened or developed to take into account different knowledge bases.

However, the confusion over the general purpose of some of the teaching that took place on this programme was not only confined to international students. While students understood that there was an expectation of autonomy, many of them expanded this expectation beyond the thinking and learning they were doing into the need to work out the purpose of some of the teaching they encountered and the tasks they were set to complete. There was repeated reference to 'having to work it out for yourself', as below:

- *a lot of the lectures are just for interest. So, this is what I find strange about the course because at undergrad all the lectures that I had were then assessed with an exam at the end whereas here it feels like we go to them out of an interest. I've never had to look back at any of the notes that I've made because the assignment then is on a really specific, like maybe one line in a lecture will be linked and then you have to just work it out yourself. (FS2)*

All the students interviewed from this site expressed some level of confusion and, at times, frustration, with the lack of clarity around task instructions, assessments or lecture content. Those who would be classed as 'home' students explained this as an expectation around working independently and had the resources to eventually 'work it out for themselves'. Mai, and other international students like her, concluded that they were 'lacking' in some way, usually both linguistically and in terms of knowledge, and therefore did not feel that they had the resources to overcome the hurdles on their own. At this point, then, is feels necessary to question what elements of a TPG curriculum should push students towards independent action, and where it might be more helpful to provide greater clarity and guidance.

This confusion from students over the purpose of certain activities and tasks crossed all Case Study sites and was repeated across different levels of the curriculum. There was confusion over the overall purpose of the programme:

- *I have my own expectations of the, since this was an MA level, that would be more practical work I would be given throughout the semester, but I mean that's the kind of impression that I received from the, from all the information that I received on the internet ... (C1)*
- *I was thinking it might have been a bit more practical than my actual degree but it wasn't, and I've got to read a lot more again but I just feel like it's better than my actual degree. (C2)*
- *It was not what I expected but I think it's better than I had expected because when I chose the new media, because my background was in information systems so I thought that it would be more practical and maybe something related to technology, but what I found that it is more theoretical. And I find it interesting. (6X)*

There was also confusion over expectations around essay writing:

- *Actually, I wondered that what's the purpose for this essay. Why do we write these essays? Because I don't know in China but here when we write the essay, we should collect many, many references to support our argument but I think there is not our ideas, just other people's. So, I'm a bit confused about this. (6Y)*

It is clear, then, that more needs to be done in terms of communicating with, explaining and supporting students to see why they are required to fulfil certain tasks, to engage them fully in their own learning and to enable them to make stronger links between content, assessment and practical applications.

So far, then, it seems that all the student participants encountered similar difficulties in terms of interpreting expectations of teachers and establishing clarity around the purpose of content taught and tasks set. The advantage that 'home' students had was not necessarily a better knowledge base or more nuanced linguistic understanding, but simply that they did not have to take this possibility into account. I will discuss the impact of social and cultural capital on language and learning in more detail in Chapter 6. It is clear, though, that international students (and often teachers who worked with them) viewed themselves as being the problem, as being in deficit in terms of language and knowledge and this then became the reasons for not understanding the purpose of what they were learning.

The impact that this view has on how the purpose of TPG programmes is conceived when the vast majority of students are, not only international but more specifically Chinese, is clear from one of the teaching participants:

- *if I'm being completely honest. And it's not unique to this university. Every university's done it, you know. A Chinese student with an MA in their hand, how much does that really mean? I wouldn't assume anything on the basis of that. (M4)*

This is a strong comment, but does draw attention to the key question being asked in this chapter: What does having a Masters degree from the United Kingdom mean? Who is the curriculum written for? Who are our students? The same teacher continued to ponder this question:

- *how do you conceptualise the difference between Masters and BA level study is a huge issue I mean, you know, and if I look at dissertations between the two levels there are times when I question whether the MA dissertation really is at a higher level in terms of the understanding and skills developed. (M4)*

In Arts, Humanities and Social Science disciplines, the knowledge base is more fractured and contested. The epistemological emphasis on 'knowers' rather than knowledge creates a more eclectic foundational knowledge base and a School cannot make assumptions around core knowledge to be built on in the way that a STEM discipline feels able to do. However, assumptions are made around key academic skills or attributes, in terms of an ability to construct or develop an argument around a theoretical framework. It is here that the implicit purpose of a Masters degree within this field becomes clear:

- *They certainly struggle with the hard-core theory. It's very demanding material and would be for English students as well but they get through it and they manage. (M9)*

In order to support 'them' (i.e. Chinese students) to manage, many teachers do make changes to their practice, but this is seen to be to the detriment of achieving the conceived purpose of a TPG programme:

- *in kind of the superficial sense that one is constantly changing module content and teaching methods to try to be inclusive to everybody who's in the programme. So, what that does do inevitably is pull down the, perhaps the sophistication of discussions in groups that are substantially struggling with basic language. (M5)*

Clearly, TPG programmes should offer challenge, and need to include 'hard-core theory'. It is also 'the crucial reality that some knowledge is *in fact* better than others' (Moore, 2007: 34); however, much of the written curriculum in UK HEIs does remain Euro-centric, as does the approach to learning and teaching (see Turner (2011) for a more detailed discussion of this). I will suggest throughout, but particularly in Chapters 7 and 8, that much more needs to be, and can be done, to enable all students to participate and access the knowledge chosen, but also that it is necessary to question our knowledge choices and become far more open to different world-views within the planned curriculum and our assessment practices. The burden should not always be on students, particularly those who are more likely to be viewed as outsiders, to engage with and be socialised into current norms; education should be transformative in both directions; teachers should learn from and be transformed by our students as much as students learn from teachers.

It is also possible to question whether a sophisticated and eloquent engagement with theory is the key factor to a student developing the 'qualities and transferable skills for employment' as outlined by the Quality Assurance Agency:

(1) The exercise of initiative and personal responsibility.
(2) Decision making in complex and unpredictable situations.
(3) The independent learning ability required for continuing professional development.

(QAA, 2012: 16, in Brown, 2015: 175)

Perhaps by widening our understanding of the purpose of a TPG curriculum beyond the written and planned learning activities of any particular programme and taking into account the entire experience that a holistic view of the University as curriculum entails, the value of a 'Chinese student with an MA in their hand' becomes clearer, as hinted at here:

- *It's creative intellect that we're looking for at Masters level and I suppose I'm a bit more fluid about what that means; 'I think the lesson perhaps is not to care so much about the intellectual outcome because a Masters' degree is much more than that'. (M10)*

The tension remains, however, between this conceptualisation of the curriculum, and the deeply rooted commitment that academics have, both in terms of their research and their teaching, to the 'intellectual outcome', i.e. the creation and development of a disciplinary knowledge base:

- *It is important to remember we are a Russell group university, so are research led, so we should not be changing what we do to suit what the customer wants, we can't just cater for a very particular international student cohort, which may change in makeup in a couple of years anyway. (M1)*

This comment suggests a resistance from academic teachers to the morphogenic impact of the interactions of an increasingly diverse population with the institutional curriculum. These interactions are not only causing a shift in understanding of the curriculum but are also seen to have an impact on the identities of those involved.

Changing Identities

Defining identity

If one of the purposes of education is understood to be transformative (Meyer & Land, 2003), that students' identities change and develop over their period of study is hardly surprising. Those engaged in providing this education do so with less of an expectation of having to face questions and reflect on their identity. Yet, as the embodiment of an institution being put under strain by a changing population and rapidly shifting landscape, it should hardly be surprising that teachers also find themselves either resisting or undergoing uncomfortable 'elaboration' (Archer, 1995).

Identity is a contested concept and is defined within a number of theoretical and research paradigms. I do not aim here to provide a thorough analysis of identity theory, but rather consider identity as a heuristic against which to map some of the comments, experiences and emotions that were experienced by my participants.

Identity can be understood 'according to what is considered essential to a particular person, type of person or group' (Moran, 2018: 5), it is then both personal and social. Both Applied Linguistics and Critical Realism suggest that identity is only really highlighted during a period of change and that it is when these essential particularities, normally taken for granted, come under some kind of threat or encounter moments of pressure that individuals and social groups begin to question who they are. Therefore, identity 'only becomes an issue when it is in crisis, when something assumed to be fixed, coherent and stable is displaced by the experience of doubt and uncertainty' (Mercer, 1990 in Ding, 2019: 66). This displacement or change in the status quo, by calling into question who you are, requires conscious reflexivity and personal interaction with the world around us. It is this loss and re-establishment of an identity that I focus on in the following discussion.

When an expected outcome of agential choice, this can lead to a sense of empowerment or, at least acceptance of the discomfort that is part of self-questioning. In this way, personal identity can become 'an achievement. It comes only at maturity, but it is not attained by all; it can be lost, yet re-established' (Archer, 2000:10). For the student participants, this need for reflexive questioning of who they were was an expected outcome of their choice to study in the United Kingdom, as part of their move towards 'maturity'. While at times more troublesome than they had expected (see Bond (2019) for more discussion of this), the majority of participants embraced the possibilities of their identity being questioned. In this way, the data fits with the Applied Linguistics notion of personal identity as being fluid and as 'the way a person understands his or her relationship to the world, how that relationship is structured across time and space, and how the person understands possibilities for the future' (Norton, 2016: 476). Thus, identity is linked to investment and individuals are able to work towards a future imagined identity (a more ideal self) through concerted investment (Norton Peirce, 1995).

However, it is also possible to take a different view of identity, as emerging (and being lost) through a moment of doubt or crisis that is not a result of personal or agential choice, but from external imposition. As I examined my data, whilst students seemed to express their identity more in terms of Norton's idea of a 'future imagined identity' (see later in the chapter); disciplinary teachers seemed to be experiencing an 'assault' on who they felt they were or should be. For these participants, the impact of changes wrought by HE internationalisation strategies, viewed as part of a neoliberal agenda, can be seen to be causing a sense of displacement and loss. As Ding (2019b: 67) argues, 'neoliberalism profoundly impacts academic identity' and is experienced by academics as an 'assault on the professions'. In other words, the identities of the staff participants (particularly from Case Study 2) were being brought into question by the internationalised and neoliberal changes to their environment, their institution and their practices over which they felt they had little control.

I have, then, taken the Critical Realist view of identity as being highlighted through change and requiring individual reflexivity and interaction with the social, natural and practical realities in which they are engaged (Archer, 2000: 9).The more optimistic placing of self within 'possibilities for the future' focuses more on individual choices around change, where it is understood that agency and autonomy can be exerted. The identity shifts experienced through 'doubt and uncertainty' suggest change being due to external forces over which the individual feels they have little agential control. The data I present and consider in the next section uses this understanding of identity as a heuristic for discussion.

Identity in the curriculum: Teachers

The theme of identity really only emerged amongst teacher–participants in sites one and two. This can possibly be explained as being a result of disciplinary preoccupations, however, I suggest that the difference lies in the extent to which staff felt they were being forced to work outside a comfortable space and within a context they were unprepared for.

In site 3 (STEM), then, although there was no reference to identity in any of the data, my notes and observations suggested that all teaching participants viewed themselves very strongly as researchers, who used their disciplinary expertise to teach students how to become researchers in the same field. There was an air of confidence around what they did, and very little questioning of their practices. This does not mean to say that care was not given to teaching and learning practices, or to students. It was. However, when a student was identified as struggling, this was not then seen as the responsibility of the academic teacher but was passed on to 'support' services and systems. This site was firmly entrenched in a 'hard pure' discipline (Becher, 1994), viewing itself as well established and academically rigorous, building cumulative, explanatory knowledge. Its student cohort remained relatively small and was not yet challenging traditional practices.

In site 2 (AHC), however, the teaching identity seemed to be much more separated from the researcher identity. Although there was frequent reference to the 'Russell Group' (i.e. research intensive) nature of the institution, there did not seem to be a strong suggestion that teaching was about training students to become researchers. More it was used as a reason to push against any suggestion of 'dumbing down' or lack of rigor that was collectively seen to be a possible consequence of cohort diversification:

- *We are rigorous but there is often a frustration around the classroom experience we can provide with such a diverse cohort who have very different needs. (M1)*

Teachers in this School had a very strong sense of who they were as a *teacher* (as opposed to a researcher) and what their educational function should be, describing themselves in the following ways: *a reflective practitioner; a performative kind of teacher; we're in the business of producing people who will inspire the world around them.*

There was a strong value placed on this aspect of their work; a sense that their purpose was to develop '*future inspiring thinkers*'. All of those interviewed had put thought into their approach to teaching and learning; a number mentioning literature on teaching and learning in HE, particularly in connection with working with international or Chinese students. Discussion and sharing of pedagogical practice were evident across the School; one example being the collaborative production of a handbook for Teaching Assistants on how to manage seminar teaching. There was, however, a strong resistance to teaching development programmes: '*academics don't like to be trained*'; '*I did a ghastly training certificate*', with a preference for learning in-house from peers, although there did not seem to be much time given to this peer-learning.

The strength of this vocalised and shared, experience-based pedagogy built a collective identity around good pedagogical practice in the site. However, across the range of participants, there was also a sense that this was being called into question, and that teachers were finding their usual pedagogical approaches were no longer effective. One of the suggested reasons for this were increasing student numbers:

- *The best way of improving the classroom experience would be through running smaller classes. There is a standard size set by the University, but we can make a decision on size to fit our needs – I could have smaller groups, but I can't take up more of my time to double up classes when I have all the other things I need to do in terms of, for example, research agendas this being a research-intensive university with research led teaching at the core. (M1)*
- *It's balancing the reality of student numbers against research findings that small group teaching works best. We need to be flexible. (M1)*

Here, the tensions between the differing pulls of academic life begin to emerge – the need to make choices between disciplinary research and providing what you believe would be the best learning experience for your students requires decisions around which aspect of work you identify yourself with the most.

It is clear that as a School, this site was struggling, not only because the number of students studying on their TPG programmes had rapidly increased, but also because this increase in numbers was due mainly to an increase in students from China. The two working in combination together was leading to a School wide questioning of practice. While these responses are individual and at times deeply personal, they build the impression of a School struggling to come to terms with a clash of educational approaches:

- *So I am personally, really disappointed if I think people haven't understood me. So there's a kind of, it's a personal thing, I mean I actually go away and feel a bit miserable about it when those things happen. (M7)*
- *It doesn't gel with what little I know about the Chinese education system ... I'm more than slightly resentful that I'm being placed in this position, like some school teacher, and therefore, I end up infantilising them by shouting ... not shouting at them but telling them to shush. I mean it feels like I'm at primary school or something. That's been really strange and I don't understand what's going on. (M9)*

This confusion is compounded when teachers have observed Chinese student interactions in China, and can see that the difference is not that they are Chinese but the context within which they are studying:

- *when I went to China ... those students were exactly the same as ours. They were chatting; they were checking their phone; some of them were answering questions; moderately engaged; moderately not engaged. They were not shy wallflowers at all. That was the first time I went and I just left there thinking, 'What are we doing wrong? Why do they shut up when they come to the UK?' There was no difference, no difference, and so that's another question that preoccupies me a little. (M10)*

Thus, there was a constant questioning at both a personal and site (but not always inter-personal) level around approaches to teaching and how to manage the differences. Concerns were expressed around the experience all students were having and the impact this had on the level of education they received. Some, although by no means all, teachers suggested there was a deepening rift between Chinese and 'home' or indeed other international students:

- *not that there's direct conflict between them, there's generally good collegial support, you know, and understanding between them but I've seen especially over the past three years much more vocal concern expressed to myself and colleagues that this should not be happening that students are not getting the education that they expected or that they feel they deserve because of the extreme difference in ability within the class and students don't see it as fair. (M5)*

Note here the use of the word 'ability' – suggesting some students were operating in deficit. This is a theme I will return to later. What is clear though is that the Chinese students are viewed as the cause of a problem. One reaction to this problem was that all participants questioned their current identities as teachers and the practices that went with this. For some, this questioning was leading to a consideration of changes to their own practice, although most seemed unsure as to what these changes should be. In fact, a number of participants commented on how they hoped I was going to be able to tell them what needed to be done as a result of my research. In this way they were attempting to maintain their self-proclaimed identity as a reflective practitioner. Others put up more resistance to the problem and felt that the issue did not lie with their approach but with the students who were not really fit to study on their programme. One teacher reported that:

- *We've had debates in this school to the extent that some tutors have said, 'What we need to do is separate the Chinese students into a different seminar group', and some have done that. (M6)*

This comment is quite shocking if taken at face value and needs further unpacking. It came not long after a student–staff forum that I attended, at which M6 was also present. The subject was raised by a student representative from a programme that M6 did not teach on. There was quite a lengthy discussion around the seminar groupings across different programmes. From this, a number of points emerged that add clarity to the comment above.

Firstly, a few programmes had such a large number of Chinese students on them that in some cases it was not really possible to create seminar groups that were not entirely Chinese. Where is was logistically possible, decisions then had to be made around the possible social isolation of one individual student from a different language background in each group. There was also confusion and disagreement from the students as to whether a student of any other nationality in a group would be preferable, or whether groupings should also ensure at least one 'home' student in each group where possible. This itself leads to the question of why this would be seen as necessary, placing greater status and authority on this one 'home' student than on all the others. Furthermore, one tutor reported having discussed this issue with his students and had worked on a request from the Chinese students who had said they would prefer to be in an all Chinese group. Other Chinese students in the forum disagreed with this position. The final difficulty around this question was an administrative one, where when seminar groups are created alphabetically from a spreadsheet, the final groupings are highly likely to be entirely Chinese due to the Anglicised spelling of Chinese family names leading to a high frequency of names beginning with W, X, Y and Z.

In this way, a simple grouping decision can take on ideological, political, sociocultural and even imperialistic undertones, however innocent the original intent. The discussion around this again led to questioning of conflicting identities within the site.

Despite this deep questioning, there remained a powerful sense of purpose as to how teaching and learning should be approached. Although there was an understanding that this might be different to students' previous experiences, in the main, teachers still believed that the assessments they set and the expectations they had within their classrooms were 'right'. Although they were clear that greater explanation of academic practices and concepts might be necessary, they were still working within their traditional pedagogical paradigm. Thus, solutions to the 'problem' included:

- *I think you have to expose them to argumentative academics. It's really helpful to take good examples of academics having a debate and say, 'This is how we do this. This is what academics debating looks like' so that they understand that this is not an acrimonious discussion. (M8)*
- *I think you also need somebody who's well able to break down concepts into those kind of multiple levels that I mentioned, so if they're not quite getting this concept, you don't keep saying, 'Well you need to' or to reiterate the same point but notch it down one level. (M8)*

- *the way we teach and we deliver and we assess. In the UK, we think that it's very common to ask people for essays. That doesn't happen in every country. You have to explain to the students what an essay is and how you do it and what you explain in it. (M3)*

In this sense, then, an increase in student numbers and an increasingly diverse (Chinese) student population did cause teaching academics to question their position, their commitments and their identities both individually and collectively. These tensions are similar to those raised by staff at Murdoch University in Australia, as presented in Four Corners (2019). Although many provided individual examples of adjustments they have made to their teaching, this had not led to great changes in overall approach or in the curriculum as a whole. Viewed through an Academic Literacies framework, the approach remained one of induction and socialisation into the current norm rather than a move towards a transformation of practice. The cultural reality of the taught curriculum in both disciplinary sites remains powerful and deeply embedded in the ways of thinking and being of those who are currently responsible for enacting it.

Identity in the curriculum: Students

Interaction with a changing population of students is, however, causing many teachers to begin questioning the practices that become the enactment of this curriculum and there is a growing understanding that it is no longer possible to teach in the way you have always taught and learned yourself:

- *what you were as a student is not a universal norm, which I think is really important for academics because we weren't the average student in the class. We've only ever known the experience of being somebody who could do it pretty easily and that's not everybody's experience at all. (M10)*

Alongside this lies an often-contradictory view of the role and identity of the students who are causing this emerging structural elaboration. These students are seen, in turn, as a problem, operating in deficit and not really worthy of a place at the University yet also as almost heroic in the efforts they are willing to put in to achieve their goal, and worthy of both admiration and empathy.

From the teachers there was a lot of concern about the potential damage that studying and struggling through an additional language might have. This included an understanding of the change needed from studying a language at school to using a language for academic study:

- *I think students who are perhaps struggling in a seminar, who find that they're going in ordinary interactions in their lives, are still thinking, 'When I was a child, I was learning how to ask for things in a shop. In School and now, I'm asking for this and I still don't get an answer. Does this mean that my entire education, up to this point, has failed me?' which isn't the case but if there was a way to … (M8)*

The shift that a student goes through in terms of their understanding of who they are and what their place in academia is when language becomes an issue was also touched upon:

- *I think it's debilitating if you're coming to a place and you just can't take care of your own basic needs, especially if you see yourself, as they are, as an academic high flyer and you're then fumbling through interactions that you expected to carry out with ease. That's probably quite damaging. (M8)*

Whilst this, for M8, was linked primarily to levels of academic confidence and willingness to participate

- *I personally would like to see the students feeling more able with spoken English … I would love to see them feeling more confident but you can't point at something and say, 'Be confident!'. (M8)*
- *they lose confidence very, very quickly. That's something, again, which can have feedbacks into their learning because they're then not asking questions and they're not asking for clarification. (M8)*

Another teacher had a starker message that takes things beyond a decrease in confidence to real health issues:

- *There are rising mental health issues. Some of these students should not be here, so far from home in an unfamiliar environment. (M1)*

This message is becoming increasingly common across the sector, (see, for example, Haber & Griffiths, 2017; Forbes-Mewett & Sawyer, 2016; UKCISA, 2019), with growing concern around the impact on mental health that comes with unsupported study abroad.

None of the student participants involved in my study reported extreme deterioration in their mental health. They did, however, report facing difficulties in both academic study and social life that echoed the comments of M8 above. A lack of confidence in speaking up in class was reported by almost all EAL students, as was some level of confusion over how to communicate with 'locals' in an appropriate manner. In terms of the impact of the TPG curriculum on their identity, student–participants did report shifts in thinking that they found uncomfortable. However, for these students, this was an expected and to an extent, hoped for, outcome of their studies:

- *For me I think it makes me understand what I am now – my current identity and my future identity, what I want to be, like the ideal me. And how the university is pushing me to change some of my attributes to change into that person. (5B)*

They were able to make direct connections with the planned curriculum and the tasks they were given within it and the way they were being encouraged to change and develop the way they thought:

- *the assessment that they gave us is one at a philosophical level, so not really practical so it's like, you would be expected to read and then kind of ... but there's no real right or wrong and the question is there just to guide you. (C1)*
- *In order to finish this essay, I did a lot of reading, but I have to say I have more questions than before. I think what I thought before is too simplified. (C3)*
- *The value of being a student and the open resources for you to transform or transfer to new fields or new aspects. (5B)*

Many participants felt that their way of thinking was changing, but were also aware that they were still deeply within the process of change, and that they would not be able to identify the ways in which the curriculum had impacted on their identity until they had distance from it:

- *But I think going along you use more than you, you learn more than you think and then things become easier. You kind of don't really notice ... so I think we are learning a lot but it's like a crash course that we have to teach ourselves. (FS2)*
- *Before this summer, I'm a shy student and I wouldn't talk. This summer, I try to talk with my friends. It's a different me, but it's better, it's progress. I won't go back. I will continue to be an outstanding student'. (8N)*
- *I think we are changing but we don't know the results. (6Y)*

Interaction with the curriculum, both planned and hidden, and with each other, had a clear impact on the identity of both teachers and students in this study. It led to a questioning of purposes and practices and a struggle to understand how best to approach the disciplinary content and how to communicate around this knowledge with each other. Key aspects of the pressures placed on identities that had formerly been viewed as fixed and stable are connected to time, trust and the ability to make agential choices. I will now explore how these three themes interacted with the participants view of the TPG curriculum.

Temporality

Concepts of time

If the questioning and changing of identity is accepted as a key issue for both teachers and students as they navigate an internationalised TPG curriculum, and it is also accepted that this identity is brought into question as a result of social and structural interactions, it is important to also consider the temporal nature of these identity shifts. Temporality is integral to the critical realist, morphostatic/morphogenic approach and is 'contained in its first axiom, that structure necessarily predates the action(s) which transform it' (Archer, 1998: 359). Taylor (2018: 295) has argued that there are increasing 'temporal tensions experienced by university academics' resulting from the 'timeless' impact of digitisation (Castells, 1996) and that conceptions and measurements of time are bound up with policy and politics rather than considerations of learning. Therefore, the normative measures of time within a UK HEI are built around workload models and key points in the calendar where time is measured in terms of quantity rather than quality. In *The Slow Professor*, Berg and Seeber (2016) also highlight the different types of time that need to be taken over different academic activities, suggesting that time within the HE context has become over-accelerated and has therefore removed a sense of belonging to a community and reduced autonomy (autonomy itself also being temporal). In this sense, it 'is learning as *being*, as much as becoming' (Taylor, 2018: 300) that takes on significance, where learning becomes embodied and part of an individual and community identity. This development of a learning being takes time that cannot be measured in calendar dates and requires, as Taylor argues, not a moving forward through time, but a staying put and a presence. In Applied Linguistics, Kramsch (2009) has made a similar point, that time is not a fixed measurement, and that language learning specifically requires more than a cognitive engagement with form. To learn a language requires viewing 'language as a living form, experienced and remembered bodily' (Kramsch, 2009: 191); however, 'the time of the body is different from the time of the mind. It is characterized by slow maturation, repetitive retracing of paths, rhythms, rituals … The body likes to re-member, re-thread, re-cognize. The time of the body is conservative' (Kramsch, 2009: 202).

The current UK University TPG curriculum is built on traditional, Western concepts of time, and time allocation for activities based on a traditional student 'norm' that no longer (if it ever did) exists. Therefore, the time needed for language development, contextual and cultural understanding is not planned for within the current, linear schedule. In this way, the 'disappearance of collective rhythms and ensuing fragmentation of social networks which become increasingly difficult to align … add to the compression of time' (Taylor, 2018: 7). A year, for many students, does not allow them either to 'become' or to 'be' and the disruption this brings also creates very real and tangible consequences to the academic calendar (for example through resits and extensions), which in turn disrupts the rhythms and structures of the people and the institution.

Time in the curriculum

A perceived lack of time to achieve everything desired or required of both teacher and student at TPG level is an unsurprising theme given that a full-time programme spans over just one year. What is noteworthy are the differences in the use of, and vocalised demands on, time across the Case Studies.

A number of AHC tutors expressed a preference for teaching UG students because it allowed them time to get to know and understand their students, whereas on TPG programmes, the amount of contact time and numbers of students involved prevented the development of the more inter-personal teacher/student relationship they claimed to prefer. This preference can be seen to link quite strongly to their own professional (and to some extent personal) identities.

While on the programme the students from the same site also felt that a year was not enough to develop both the academic and language skills required to understand their subject fully:

- *we are not yet able to question the teacher and the texts; it's our habit. We need more time to develop this. (8M)*

However, a number of these same students also maintained that one of the main reasons they chose to study in the United Kingdom was because the Masters degree only takes one year to complete (in contrast to two elsewhere):

- *I think studying in Britain, only for one year is really short and but if I go to Japan or America, I need to spend two or three years and I think it's not very good for me ... This is my first reason, the most important reason why I came to Britain. (6Z)*

This was linked to a sense of urgency around being able to enter the job market and despite a recognition of the difficulties they encountered by only being on a programme for a year, they continued to argue that a year was better than two for them to have spent on their studies.

This instrumental approach and sense of being time poor is reinforced by comments made around the length of time and amount of formal and informal classes they had on their programme. Classes attached to credit-bearing modules were highly valued; a number of comments were made around a desire for the length of seminars to be extended from one to two hours. In fact, teachers in the AHC School (Site 2) had, on some programmes, already extended class length for 2016/17 to provide extra time for explanation and discussion. In contrast, the Study Groups created by the same School to encourage peer support and teaching outside the formal classroom, although seen as theoretically useful, were engaged in to varying (lesser) degrees. The argument was that they would be better with a teacher present to make sure the students had understood correctly. Furthermore, in feedback on the insessional classes (timetabled by the School, but taught by EAP teachers and carrying no credit), students agreed that the classes were beneficial, but argued that they would be more so if they were connected to credit, arguing that they would be better attended and taken more seriously. Students were making strategic choices as to where to spend their limited time, but their choices were largely biased towards teacher/knower fronted sessions with little value placed on extra-curricular activities.

In the STEM site, there was little questioning of the appropriacy of taking a one-year full-time Masters course. The first Semester was comfortably viewed as a bridge from UG study; the second Semester a bridge into research. The programme was perceived as intensive, but manageable. Students appeared to create their own study groups in the computer cluster (although these generally seemed to unconsciously exclude international students) and accepted the decreasing amount of contact time available to them throughout the year. Here then, Mai was in the minority in terms of feeling a need for more time to acclimatise to her disciplinary culture and context.

For those students in the AHC Site who felt that they struggled with language, the time taken to read texts was also frequently seen as an issue and understood as such by their teachers. This limited the volume students were able to read prior to formal teaching sessions, and thus the ability of teachers to work with their students to extend and stretch thinking beyond the basics.

- *I think you have to manage your time in advance to maybe have more time for the reading and to think about the ideas and focus on the whole structures and how clearly, logical, structure to write your essay and … yes. (6Y)*

In the STEM School (Site 3), most students also expressed difficulties with reading. Here though, the difficulty was not only with highly specialised information, but also at times the overly complex instructions provided by their teachers. However, they also appeared to accept the need to do this as part of their learning process.

- *It takes ages [working out what to do]. So often there'll be a set of guidelines on the VLE, which is useful. It's never just a title ever. It, often we're directed to a website or a few different things to read or whatever. And then after that we kind of just have to sit down and look through them and follow the links to other things and I guess we had quite a lot of practice in looking at papers and following back to see where those results came from so references, following those references and things like that. (FS2)*

There was little suggestion, other than from Mai, that learning in class was prohibited by lack of, or extended as a result of extensive reading. Reading beyond core introductory texts was viewed as being for personal extension rather than group discussion.

Across all interviews with teachers in the AHC School (Site 2), the demands on time created by the marking of 'inelegant' scripts written by low proficiency EAL speakers was emphasised (see Chapter 5 for further discussion on language and assessment practices). Tutors described in some detail their (usually unsuccessful) attempts to keep to an allotted amount of time for marking individual scripts. The encroachment of this marking load, which was compounded by the large numbers of students involved, was seen as having a potentially negative impact on the other work required when on a Teaching and Research contract at a Russell Group University, namely research. Tutors did, however, take the time to provide clear feedback as there was a strong sense amongst the participants that students needed this effort in order to be able to develop their understanding of what was required of them in future pieces of writing. This was echoed by the students and suggests a School-wide understanding that TPG students require this clarity and support in order to develop academically. This feedback, and the time spent on it, was seen as an important part of the taught curriculum.

In the STEM site, assessment and marking also obviously took up tutor time. However, working with international or EAL students was not seen as obviously overburdensome in this work. Mai was seen as an anomaly in terms of linguistic proficiency in the School and was very quickly highlighted as a potential 'fail', thus triggering requests for further language support. This immediately removed the duty of care, and therefore the burden of time spent, from the content teacher and placed it on adjunct Student Support services. Within this School, TPG students seemed to be initially supported but increasingly left to develop their academic understandings on their own, rather than through tutor feedback. While there was no reported lack of time for student education or assessment, time was carefully guarded, and staff were generally perceived by students as being unavailable outside allotted teaching and office hours. All extra-curricular support and consideration was handed over to other University services and those with specific roles focusing on Student Education.

Perception of time, then, appears to connect with both disciplinary teacher identity, the student context in which learning is taking place, and the expectations and previous study and language backgrounds of students. In the STEM site (3), with fewer international students and a strong research-focused approach to work, time was protected by academic staff for their own research, with students being expected to rely on support services and each other for extra help. Students accepted the necessity of spending a lot of time developing their own understanding and working alone and did not, therefore, place a strain on the system as it was. In the AHC site (2), time for research was being eroded by increasing student numbers and by the time it took to enable these students to understand the culture and processes of learning within a new environment. There was a general recognition that there was not enough time to teach or guide the large number of TPG students who were enrolling on programmes in this site in the way that staff would have liked to. In this way, the change in cohort not only impacted their understanding of themselves as teachers, the time they had to give to individuals, but also the trust and confidence they had in the curriculum and the 'product' being offered to students.

The Importance of Trust and Emotion

Following on from Becher and Trowler's work on *Academic Tribes and their Territories* (2001), which found that academics developed their academic identity both through the institutionalised structure of their discipline and via the social network of their peers, Roxå and Mårtensson (2009, 2016) considered whether those academics who were also involved in HE teaching went through a similar process of identity development and learning. They argue that teachers create small but 'significant networks' or 'microcultures' that are made up of 'significant others'. These networks are developed within disciplinary boundaries and take the form of conversations that are informal in nature – on the bus, over coffee, noticing a colleague's door is open and popping in for a chat. It is through these significant conversations that 'teachers allow themselves to be influenced to such an extent that they develop, or even sometimes drastically change, their personal understanding of teaching and learning' (Roxå & Mårtensson, 2009: 548).

There are three key features to these conversations that make them significant. The conversations are intellectually intriguing, they are private, and they are trustful. This trust is based on mutual respect, on an understanding that anything that is said in the conversation will remain private, on a belief that it is possible to learn from the conversation – that the 'significant other' has expertise to offer or that the conversation will lead to a shared shift in perception. Roxå and Mårtensson have suggested that these networks are strongly linked to levels of student engagement within a School but also that they 'are vital for understanding how academic teaching cultures display a remarkable degree of resilience towards external pressure' (2016: 134). It is through these networks that teachers gain intellectual support and development around their teaching practices, but possibly more importantly, meet each other's emotional needs, support each other to make autonomous decisions that may not always fit within more official discourse.

It is, therefore, necessary to consider trust in conjunction with the emotional as well as the cognitive elements of learning and teaching. I revisit emotional support as an element of social capital further on, in Chapter 6. Increasingly, emotion is being viewed as an integral element of an individual's identity and sense of agency:

> teachers use emotions as a resource in talking about their agency in terms of their response to others: they orient to and engage in collaborative acts of affective stancetaking … to construct wider, shared discourses about their ideals, constraints, and answerability as teachers, and in so doing co-author themselves as a fundamental part of their agency. (White, 2018: 595)

Emotion, like language, cannot be separated from thought, so is also linked to *knowing*. Following Vygotsky, Lantolf and Swain consider the concept of *perezhivani* 'as an emotional experience that motivates thinking and that thinking in turn always implicates an emotional reaction to objects and events' (2019: 529). Just as identity only becomes an issue when in crisis, emotions are heightened and highlighted when normal thought processes are interrupted. These emotions can be both negative – around loss of control, and concerns over accountability at work, for example, or positive, as with trust.

Trust in the curriculum: Teachers

Within my own investigation, there was evidence of this trusting relationship, built around microcultures within the two disciplinary case study sites. In line with Roxå and Mårtensson's findings, this was stronger within the AHC site, with individuals mentioning support received from colleagues. However, in both sites, much of this trust in terms of student education practice was placed on those who held specific student education-focused roles within the School. There was little evidence of staff having developed sustained microcultures, built over time and based on privacy, intellectual development and trust. In fact, within the interviews I conducted there were occasions when some participants took advantage of the open ended format of our conversation to engage with me as a 'significant other', as someone who was asking them about complex issues in their teaching practice and allowing them to express their ideas and check the validity of some of their approaches. While participants were aware of the official support and procedures, there appeared to be a vacuum in terms of these small significant networks that are vital to the emotional support and professional teaching development of staff.

Without these strong microcultures of support, teachers in this site were feeling embattled and had lost trust in a system that seemed to focus on 'internationalising' an institution for financial gain rather than to develop a culturally diverse learning context. There was a lack of trust in the admissions process and in the wider institution that seemed to be pushing for increased student numbers without obviously providing the support for teachers to manage this within their usual pedagogical practices.

> *I just have a feeling that we've kind of lost a bit of quality control around MA's. I've just got a feeling that there is a general semi-stated view that MA's are not really as important as everything else we do. (M7)*

A number of teachers in the AHC site felt that there was a clear difference between expectations around and the value placed on Masters level programming and the work done at both undergraduate and PhD level. There seemed to be a constant pull between the maintenance of academic rigour and credibility and the reality of the TPG teaching context within the UK Higher Education landscape. A direct link was made between the surge in international student numbers at TPG level and this erosion of quality. There was frequent re-affirmation of the strength and theoretical nature of the academic offer made by the School:

- *we get some of the strongest applicants. (M3)*
- *we're a very research-based institution so we do a lot of theory and quite academic stuff. (M2)*

This was then offset by a sense that this offer was weakened by increasing student numbers and a changing cohort of students:

- *Because of the numbers that we're getting in, because of lack of qualification of some of the people for the courses that they're coming for, because I think there is a problem of balance within the courses in terms of the cultural backgrounds of students, which is very different from saying we don't want different cultural backgrounds. It's about balance. And it's about, I think also a tendency to kind of squeeze more financially out of those than we ever think of doing around undergraduate and PhD teaching. (M7)*

However, there was a strong expression of localised trust; of confidence in their own School's ability to manage in difficult circumstances and of the strength and rigour of their discipline to continue to maintain quality teaching and learning. A great deal of reliance and trust was placed on those people who held specific roles in Student Support and Student Education, both administrative and academic.

- *I think we've got a fantastic team who work on student education here, I don't know what other schools are like but I'm really impressed ... the people who do it, I'm immensely impressed by. So, they give a lot of thought to it. (M7)*

In the STEM site, this reliance was in terms of 'these are the people to go to when there is a problem or when I have a question; they are the people who are fully responsible for considering anything in relation to student education which is extra to the delivery of the modular content'. These specific people were very engaged and dedicated to student education but this culture of engagement was not endemic and they had developed their own microcultures with peers who were generally outside rather than inside their School. There was little evidence of significant cross-School or intra-School discussion of issues relating to student education in general, whilst language and the linguistic burden placed on students by the disciplinary requirements was seen as, more or less, a non-issue. When language was flagged as being problematic, as with Mai, this was viewed as a support need, disconnected from the content teaching and requirements. Thus, again, trust was placed outside, in the EAP unit, to provide the required language support, largely in isolation to the content teaching. Here, then, there was strong trust in the systems in place and reliance that these would work, allowing each person to get on with their own particular job, very much in isolation of each other. There was also clear trust in students' academic abilities and their ability to work autonomously, without extra intervention from their teachers. There was trust in their 'product', in the subject they were teaching, the quality of their offer, the research training and knowledge it provided.

In the AHC site, although there was an equally strong reliance on, and trust in, those with student education roles, this was in terms of these staff members working across the School to develop shared understanding and practices as well as taking on the burden of dealing with 'issues'. Within every interview conducted, there was mention of colleagues who had provided support or ideas for teaching and learning. There was a sense of a cross-School dialogue and debate taking place in connection to how, as a School, they were going to work with and adapt to the changing (international) TPG cohort. Although this School worked closely with the EAP unit and made use of extra EAP development provision for its students, this was not seen as being fully outside the remit of content teachers.

Much of this change in practice was still, though, presented as being imposed upon individuals, leading to a loss of autonomy. The extent of this feeling could, to an extent, be explained as being due to a lack of the 'autonomy support' that Roxå and Mårtensson (2009) suggest develops through significant conversations and friendship support. However, it was also clear that, whilst trust of colleagues was strong, there was an increasing loss of trust in the macro level of university policies. The power of the neoliberal university, driven by economic forces, was eroding trust in an institution that seemed to place decreasing value on the people who embodied it.

Trust in the curriculum: Students

Trust in students in this site was on a broader scale and was framed, by teaching staff, in terms of their ability to cope with living and studying in a different language and culture, rather than simply focusing on their ability to take responsibility for their own academic learning:

- *The fact that they are juggling with so many different things at once. The new country; the financial pressure; the obligation; the desire to have a good time; the desire to travel; the need to do the work; the nervousness about doing it. There are so many different things going on in their minds and I think they have to be pretty tough characters to get through that. But many of them do and that's fantastic. (M10)*

While teachers in the AHC site were losing trust in their ability to offer a quality TPG education as a result of growing student numbers and the range of entry qualifications, language proficiency and academic expectations of the students they were working with, this lack of confidence in the value of their programmes was not mirrored by the student–participants. However, the locus of trust does make it clear why teachers from this discipline felt intense time pressure and experienced an increasing difficulty in being able to provide the educational experience that they felt to be valuable. It also highlights that the need for the development of supportive microcultures amongst students is as great as it is amongst teachers.

In the AHC site, the trust expressed by students was clearly focused almost entirely on the teacher; the 'knower' rather than the 'knowledge' (Maton, 2014). Students focused on individual teachers' abilities to clarify and explain key concepts; there was much more attention paid to the speed and clarity of response to email requests, for example. Whereas in the STEM site, the students' main recourse for content or instructional clarification was each other, in AHC the most cited response from students was to wait at the end of a lecture to ask questions of the tutor and then, if there was still a lack of clarity, to email. Disappointment over allocation of dissertation supervisor, for example, was couched firstly in terms of the amount of individual support a student believed a tutor would be willing to provide and secondly in terms of the specific methodological knowledge they held.

This need for confidence and trust in individual teachers as 'knowers' seems to stem in part from a lack of trust in each other. The majority of the student–participants' peers were also international/Chinese; as such, although they did spend a lot of time discussing their studies together, this rarely seemed to result in a confident resolution. Emails to tutors were frequently sent as a collective, representing all concerns from a group of students who had failed, collectively, to understand concepts or requirements. These students, whilst providing a great deal of social peer support, felt that they lacked the cultural/knowledge capital which would allow them to support each other academically.

- *I think they are very good Tutors and supervisor, I think the main problem is maybe is from ours, so we have to try our best to learn and study the, our course, so. Because we are lazy [laughs], we are lazy … Because the reading is too difficult so I don't want to. (6Y)*

In the STEM site, trust was placed in the subject itself, in its intrinsic value and in the cutting-edge nature of the knowledge provided by this particular School. Thus, the knowledge they were gaining was seen as unique to their programme, and of value because of its particularities.

- *Nano particles, learning about them doesn't help you learn about something else so like here It's like looking at stuff that's the cutting edge that's been done but not like general principles, it's very specialist stuff. (FS5)*

Those students who had studied in other HE Institutions (whether in the United Kingdom or elsewhere) for their undergraduate degree accepted the gaps in expected knowledge as a necessary aspect of this, took responsibility for plugging them and placed trust in each other to help with this. These students had developed their own microculture of learning support.

- *I always come into uni to work. And I always go to the computer room so there are other people around. Just so that you can ask them things and bounce ideas off them and things like that. Because I think that is massively useful to be able to talk to other people. Even just like, there can just be like some little thing you don't understand and they can just help you with that or you get ideas off them. (FS5)*
- *I try to stay here when I can because when I'm here I'm focused on work. There's too many other distractions at home … it is a difficult course*

and, you know, everyone struggles to some extent so there is some sort of communication on helping each other as well, which is nice. (FS3)
- *In the computer room with my friends. I've never written a paper critique before and one of, and my deadline for next week, I have a week to write a paper critique and I don't know where to start and so I was talking to my friend yesterday about that and we were just like, 'We don't know where to start'. We didn't reach many conclusions but as we go along I think it will be useful to discuss it with them. (FS2)*

This learning community, for this discipline, seemed to be an unwritten part of the curriculum. Teachers seemed to assume this sharing of not-understanding among peers would take place and enable most students to collectively learn – both inside and outside the formal learning environment. However, as it was not an explicitly named learning space or opportunity, it was one in which Mai, for example, did not participate. The social conversation that took place around this learning served to inhibit and exclude her because:

- *if I listen the English a long time maybe I have some headache, yeah. So maybe in one hour of class I just listen half an hour. (FS3)*

The most obvious person for her to choose to focus her 30 minutes of listening on is the teacher, not her peers.

Amongst all students in Site 3, less confidence was placed in the educational practices of the teachers. There was criticism over clarity of assessment instructions, marking practices, the timing of assessments and the overall planning and cohesion of modules and programmes. Yet, for the project participants, this seemed of secondary value to the knowledge available via access to the research and the researchers.

- *here it's always a bit of a stab in the dark because sometimes we're not given a huge amount of guidelines. And I think actually that results in us producing quite different work to our friends, which is just an interesting point but I haven't felt like that's been limiting my marks because I've just sort of used my common sense and then tried to work out what's important and maybe what they're trying to assess for each one and focusing on that. (FS2)*

Only the International students expressed a need to access teachers on a different level in order to gain clarity and deeper understanding of the content and the expectations placed upon them. For one international student, this access was requested early in Semester 1 and was used as a means to gain reassurance that her difficulties were not unique; she then transferred her trust to peers. For Mai, her discussion with peers was limited only to other international students, specifically one other Chinese student. She felt that the knowledge and language of the UK students was as inaccessible to her as that of her tutors and therefore it would be more productive to speak directly to the teacher, until she also lost confidence in her ability to understand here as well.

What is clear is that a TPG curriculum is developed on the basis of trust that moves in multiple directions. Students need to trust that their teachers are providing them with access to key disciplinary knowledge and that they are developing an understanding of an area that will enable them to meet their future goals. Academic teachers need to trust that the students they are working with begin their programme of study with the core attributes that will enable them to fully engage with the disciplinary knowledge base they are providing; at TPG level there is also a reliance on students being able to trust and work with their peers in order to develop and deepen their understanding of their subject beyond the classroom. It is clear that, at least in the AHC site, there were multiple breakdowns in this trust and that the lack of trust is focused on students who form a large, seemingly homogenous social group that is both different from the host institutional norm yet dominates a cohort of students. In the case of this context, students from China. This 'othering' of this group of students is enacted by both Chinese and non-Chinese students and teachers, with an increasing lack of trust in their ability to fully participate in UK academic life. This international (but non-Chinese) student explains where he places his trust in his 'professors', and where he feels the breakdown occurs:

- *One observation that I found is one, it's really hard for them [Chinese students] to learn to comprehend the text. I myself am having difficultly whether you just kind of, oh this is the part where … oh okay, it's not really relevant and that part isn't so I'm just going to focus on that, it's about synthesising the text that's being presented to you, and I think that Tutor kind of gives you a bit of a variety of readings of different difficulties to understand the concept that is being presented, although they present the same idea, but then, and I think this is where they are a Professor really, because they can pick out readings that are very different to another in terms of difficulty, in terms of understanding the readings, but after you kind of step back and then you kind of put the papers on your desk and hopefully you talk about the same thing, it's actually complementing each other, but to these students I think it's a matter of one, understanding and the second, they don't want to ask … (C1)*

This places an increasing burden on the teacher to transmit increasing amounts of information to students in order to enable them to access the curriculum and disciplinary knowledge. Many teachers feel that they are doing this in isolation, with procedural but not the developmental or emotional support that would help them to feel that they can do this with success.

Making Agential Decisions

Defining agency

Agency 'is normally meant to designate that subset of human behaviour that we wilfully direct and that is distinguished by its involvement in an intentional chain' (Porpora, 2015: 55). Agency is a concept that relates specifically to human activity and thought; it is bound up with our personal and social identities and our commitments; 'we are who we are because of what we care about' (Archer, 2000: 10). These commitments, the things that we care about, are also 'subject to continuous internal review' (2000: 12). However, this reflexive, agential process does not take place without a context or a social structure to intentionally act within or against, without something that provides the reasons for acting.

Here I highlight two conceptions of this social/structural context within which the actors of my study felt enabled or blocked in their desires to make agential choices. The first is Archer's description of 'Corporate Agency', that is the actions, choices and accountability of individuals who view themselves as representing an institution. These agential actions are not necessarily made as a result of personal commitments but come from ideals or strategies that an individual has either bought into or taken on due to their particular role within an institution. Archer argues that the corporate agent, 'transforms itself in pursuing social transformation … by inducing the elaboration of the institutional role structure' (2000: 11). I suggest that, in this particular context, by pursuing and supporting a more inclusive, internationalised culture, the university (and by association those who educate within it) is transforming itself, and therefore creating social change. However, this collective 'corporate' action has also created new roles and structures within the previous *status quo* which staff are required to assume. A new, or elaborated, social identity is derived from or assigned to the role which may or may not connect with an individual's identity. I suggest that as the imperative to act as a Corporate Agent increases and is increasingly privileged in the modern neoliberal university, opportunities for individual agency are reduced.

The second is an understanding of the TPG curriculum and the wider University as being itself human, in the sense that it is enacted by human actors. Following Bhaskar, Archer (1998: 359) suggests that where 'the relations into which people enter pre-exist the individuals who enter them, and whose activity reproduces or transforms them, so they are themselves structures'. Thus, there are normative expectations around the relationships and interactions that take place between students and teachers within a particular level of study, a context and an educational culture that interact with the social structure of the University. This structure has remained generally static for those who have been part of it for a number of years, and it has become closely intertwined with their social identity. I suggest here that the interaction with those who are not socialised into the current structure is causing rapid and complex morphogenesis and entailing a much more urgent questioning of personal and institutional commitments. For students, having made the clearly agential choice to study in the UK, this transformation was expected; for many of the teachers involved the resulting shifts led to a loss of or change in both personal and social identity and therefore a decreased sense of personal agency. While 'it is the person who arbitrates upon the relative importance of their multiple social roles and between their greedy demands' (Archer, 2000: 12), for many of the participants (both teachers and students), there was a sense that there were too many 'greedy demands' which held equal importance. It was this that reduced agential choice.

Agency in the curriculum

There was a strong connection between the extent to which participants trusted the TPG curriculum and the processes it took them through and the extent to which they felt able to make agential decisions. Whilst there was a strong sense of agency across both disciplinary Case Study sites and amongst both students and teachers, it was clear that some participants felt that this was eroded as a result of the actions of others

Students in the AHC site particularly expressed a sense of being able to act within and influence their own academic environment. Throughout the data collection period, participants expressed a developing understanding that this was a key part of their learning, and that the institution provided opportunities for them to develop this sense of agency. This was viewed as an important outcome of their M-level journey:

- *I think academic life is an opportunity to just talk about whatever you want, as long as you respect others and as long as you bring your arguments in a clear way. That's also for the other person, of course, if he wants to accept it or not. (5B)*
- *Also, I think, we have different methods of learning. In China, we are always taught by someone and you have to be pushed by a teacher. In the University ... we have different ways of learning because we have to research by ourselves and read more articles and finally, we get the conclusion and have our own opinions. It's maybe helped by a teacher*

but the teachers and tutors are not the main road in our study. I think it's different from the Chinese method. (5E)

Students focused on decisions they were making around academic processes, on how they were developing as independent and autonomous thinkers and their growing confidence in their own abilities. As with Lin, when discussing shifts in understanding, students were keen to suggest that this was a result of independent realisations and agential action. However, there was also an understanding that the support provided by teachers and that working through the feedback they provided enabled this process of self-development.

- *I feel like I know how to start and how to arrange the structure and the references. I feel like it will take less time than it did before. (5B)*
- *I realise that it's not about the amount of ideas that you put in, the amount of reflection that you give, that you can put in after you've read that, that's where the marks come in and I just got this, after I received my first assessment back, and I went okay, well I'm pretty happy with the marks but then I kind of read the comments, read the whole thing again, oh so it's not about the amount of read things, they don't care that you read 100 books, they don't care, what matters is whether your voice is in it, what do you think about all of that, are you agreed or disagreed ... (C1)*

When the support was removed and students were expected to work without clarity, they also experienced a loss of agency as they had no sense of a framework or parameters in which to work. Assessment practice in the STEM site were frequently cited as lacking clarity. Yet whilst in most cases this was viewed as a means of developing independence, and students took responsibility for deciding what was needed and finding out for themselves, in the following case the result was a sense of powerlessness and inability to act:

- *There's one particular module ..., and they had set work with actually no guidelines including even how, you know, word count or anything that, no expectations at all, which several of us had queried and asked the person who'd set the work, you know, and we got quite a sort of, you know, irritated answer back saying, 'Well, I just, you're Masters' students and I expect you to do that'. The only guideline he gave was that he wanted, it was a series of questions, you know, and it wasn't clear whether they were essay questions or what they were, were you supposed to put the references in and refer to literature or there just wasn't any guidelines for that at all. All he said was, 'I expect concise answers to the questions'. (FS4)*

It was when power was removed that individuals experienced the greatest loss of agency. For students this is experienced through a lack of clarity in assessment and uncertainty around expectations.

For their teachers, one of the problems with a TPG programme is that these students, only at an institution for a year, are never fully able to assert themselves in a meaningful way on the structures of the University:

- *And of course they go after a year and they can't follow up but that's why we need to speak up and say this as a progression and I'm not sure if it increases, that frustration, I'm not sure how we're going to manage it in the future. (M5)*

In the neoliberal university, the student, or 'customer', voice is often key to making change happen. For the teacher–participants who felt that the most negative impact of a neoliberal financially oriented internationalisation strategy was being felt at TPG level, there was frustration over the lack of opportunity to follow up on the issues some of their students were raising, and the lack of agency they had in working to manage these issues.

Despite this, teachers did still see themselves as agential at work and as having high levels of academic freedom. Within the context of this study, disciplinary teacher agency was expressed in terms of having choice over programme and module content and in being able to make changes to this where and when appropriate (albeit within the constricts of the institution wide Quality Assurance systems). Agency was at its strongest when teaching linked closely with an individual's research interests. However, there was, again, a sense of growing tension between the agential strength individuals felt they had, and should have, as academics and researchers, and the perception of the gradual erosion of this as a result of changes to university structures, policies and procedures.

This was expressed again around admissions procedures, which were seen to have a direct impact on the quality, and therefore purpose and value of TPG study:

- *ultimately what's needed most is trust in the academic professional to evaluate who's qualified to take a particular programme and who isn't and that's what's often missing and is gradually diminished year by year as that control is increasingly centralised. It's not just a University of XXX thing, it happens everywhere, but the problem is that it results in an unfair situation for academics who have to teach people increasingly who are not in, do not have the ability to learn the things that they're being asked to teach them and it's terribly unfair to students who are brought in with an expectation of being able to get a degree and not actually having the ability to do that. And my, so my preference, my request on my wish list is that we actually be listened to when we say that our standards of admission actually need to be higher if the degree is going to have value and actually work in terms of being able to be delivered.* (M5)

This involvement in admissions procedures seemed to be a procedural difference between the two content sites of study. Possibly due to the smaller number of applicants, the teachers in the STEM site still seemed to believe that they had a good deal of autonomy and responsibility for decision making around admissions, in contrast to the frustrations reported by M5 and others.

The erosion of academic agency is seen to impact negatively on all aspects of teaching and learning, from admissions through to final assessment and award of degree:

- *Because we've got the criteria that we're all working to and quite often we have sessions where we do this and see if we're working to the criteria. Then the criteria get adapted because of criteria change to the hundred scale. Then there's always like some divergence, but basically nobody would fail it. It's such an imprecise thing and then we have to pretend that it's scientific. Then we have to give feedback that they can accept the mark and we have to say .... But also, we have to say that we're all marking to the criteria, even though we do know that it's subjective to some extent even with the criteria. So, I find it quite hard ... a minefield.* (M1)

Teachers, therefore, seemed to be working to maintain their academic ideal as agential actors, providing what they currently understand as a rigorous, research-led post-graduate education, whilst being pulled by the changes to the student population towards a reconsideration and reconceptualisation of what this might mean in real terms.

Chapter Summary and Practical Lessons Learned: TPG Education

This chapter has considered the differing conceptions and experiences of teaching and studying through a taught post-graduate curriculum within a UK Higher Education institution. The curriculum is framed as a socially real institution that impacts on, and is impacted by, the individuals who enact it. This impact is seen through the lens of four key themes: identity, time, trust and agency.

I have argued that there is a lack of clarity around the purpose of a taught post-graduate programme, and that teachers who work with large numbers of international students at this level report experiencing a sense of erosion of quality around the educational experience they are able to offer. Interaction with an increasing international, and in particular Chinese, cohort of students is having an impact on the academic and teacher identity of those responsible for delivering the curriculum. There is also an impact on the time they have available to both teach as they would wish to and conduct the research that they also need to do. As a result, the trust they have in their institution, in the decisions it makes around admissions and quality are also being eroded, although not their trust in their peers and their discipline. This erosion in trust has also led to a sense that decision-making abilities had been removed and that their own agency is being depleted. In a site that is not yet experiencing a significant change to its traditional student intake, this impact on teaching staff is not yet being felt so deeply.

From a student perspective, whilst the curriculum they experienced did not necessarily match their initial expectations, for most participants it exceeded them. The majority reported a shift in identity that included both difficulties and positive shifts, but also a sense that any real change in identity could not be evaluated until their TPG experience was complete. Time was a key difficulty for students, in terms of the time taken to read and understand texts, the timing of assessments, and the time needed to develop both language and content understanding. Yet a year was still seen as the preferable amount of time to spend on TPG study. Although students generally trusted their teachers and the institution to provide them with a good education, there was a difference in the level of trust students placed on each other between different cohorts. It is here that 'othering' and 'self-othering' begins to take place, as Chinese students who were studying as a majority group failed to trust each other to provide academic or disciplinary guidance. This contrasts sharply with the trust that students from any other country, working together, placed on each other to 'work things out together'.

There are clear messages in this chapter that can be taken up, in some cases at an institutional level and in others at a more individual level in terms of practice. I begin with suggestions made by some of my participants.

- Consider the relevance of content to all your students; develop reading lists in partnership with your students:
  - *I think we've become a lot more aware of the student experience. I think we have de-Westernised some areas of the curriculum. (M8)*
  - *The, you know, content is tailored to the students. It is designed to engage them with things that are both challenging and familiar so there is a lot of Chinese media. (M4)*
  - *The first thing I check when I have a cohort is the nationality of the students. Then I prepare my classes with examples that relate to them because 90% of them are going to go back to their country to work and to be a journalist or a public relation practitioner. They're going to go back to their own society. Unless I make my teaching relevant to them ... (M3)*

(I would question some of these practices and wonder whether students who choose to study outside their own country want to be introduced to a wider range of studies and contexts, rather than just those of their own country and the United Kingdom. Perhaps an international/global approach rather than a simple UK/China binary might be more appropriate?)

- Provide more face-to-face/timetabled support. Ensure it is directly connected to credit/assessments:
  - *we made the lecture longer to accommodate so if there was any questions or so we could explain things in a little more detail so we made the lectures a two hour slot. (M2)*
  - *Study groups I think are really important. So, they debate together before they come to a seminar and that's a new thing for me. (M6)*
- Any group work that is an expected element of learning must be timetabled; if it is not then the student who is the most 'powerful' in the group (almost always the 'home' students) is able to dictate meeting times which may clash with the commitments of others – for example, if language classes are provided as a bolt-on session, international students feel forced to miss such classes to meet their group at the same time
- Be explicit about the amount of reading and what needs to be read.
- Structure seminars more carefully; don't rely on students to join in without guidance.
- Find ways to build trust amongst students to reduce reliance on the perceived 'knower'.
- Encourage and make time for teachers to talk about teaching within their own context. Make it part of their disciplinary identity work.

These are just a few suggestions (see Bond (2018) for further ideas). There are many other ways in which teaching and learning practices can work to be more inclusive of all. Developing a fully inclusive curriculum is, however, an almost impossible task. It is possible to argue that almost every teaching method can exclude one person through a desire to include someone else. What is required more than anything else is a reflexive view of practice that is open to (self-)critique and has the flexibility to change and adapt with the students. It is necessary to be continuously working *towards* inclusion rather than assuming that full inclusion can be achieved. Part of this inclusive practice is a consideration of how language impacts on our students. It is here that the role of EAP practitioners within the disciplinary curriculum and as a counterbalance (or possibly pharmakon) to the issues outlined above begins to emerge.

4 Language and the Academic Curriculum

I have so far considered the impact of and on a diverse and internationalised student population through the lens of the disciplinary curriculum. I now move to consider the interplay of (academic) language across this curriculum.

Within the field of EAP, the case for connecting language and disciplinary content knowledge is now well established (see Boden-Galvez & Ding, 2019; Hutchison & Waters, 1987; Hyland, 2002, 2016). Although discussions continue around where the line is drawn between English for General Academic Purposes (EGAP) and English for Specific Academic Purposes (ESAP), and to what degree of specificity EAP teaching needs to go, there is general agreement that developing students' knowledge of and ability to use English for academic purposes is better placed within the boundaries of their own disciplinary discourses.

Much research in the field of EAP now focuses on establishing just what these discourses and genres within a specific field look like (Gimenez, 2009; Hyland, 2002, 2004; Hyland & Tse, 2007; Nesi & Gardner, 2012; Gardner *et al.*, 2018). The focus of this research is largely on the written word and looks at published research papers. Nesi and Gardner's 2012 study of undergraduate writing genres signals a shift in focus beyond the more conventional forms of published academic work to a consideration of student work which is, after all, the target genre for most EAP students. However, Nesi and Gardner, and latterly Gardner *et al.* (2018) still describe texts as a general 'type', leaving the social practice of writing within a discipline for the practitioner to consider. Increasingly more practice-based research has begun to consider what kind of language, discourse and literacy skills students might need within specific disciplines. Thus Hartig (2017) has taken a case study approach in order to understand what connects language and legal disciplinary knowledge, focusing very much on conceptual meaning and how this meaning does or does not transfer across languages, cultures and different legal contexts. Sloan and Porter (2010) have proposed the CEM (contextualised, embedded and mapped onto an academic programme) model for engaging business academics with insessional EAP work; Wingate (2015, 2018) has also argued for academic literacy being developed through collaboration within a disciplinary community. In *Working with academic literacies: case studies towards transformative practice,* Lillis *et al.* (2015) bring together a collection of pedagogical approaches and case studies that outline how collaborations with disciplines can work to engage and support students across a range of contexts. While EAP research firmly suggests the need for a disciplinary focus, there are very few openly available teaching resources for practitioners that meet this need. Garnet have published a range of ESAP coursebooks, for example, 'English for [Medicine] in Higher Education'. However, the difficulties of developing a context-specific, disciplinary focused EAP programme and resources are painfully outlined by Pajak (2018); most currently available disciplinary resources seem to focus more on skills than language.

To date, though, it seems that the EAP/Academic Literacies community has mainly been talking to itself. There are very few published examples (although there may be many examples in practice) of the kinds of collaborative work that Wingate (2015) suggests, which enable EAP practitioners to work alongside content specialists to support students in the development of key academic attributes, language and literacies. Hyland argues that it is

> difficult to separate completely the teaching of specific skills and rhetoric from the teaching of a subject itself because what counts as convincing argument, appropriate tone, persuasive interaction, and so on, is managed for a particular audience. Students do not learn in a cultural vacuum but are judged on their use of discourses that insiders are likely to find effective and persuasive, yet also claims it seems evident that subject teachers generally lack both the expertise and desire to teach literacy skills ... Subject specialists often believe that academic discourse conventions are self-evident. (2016b: 19)

Therefore, whilst the field of EAP may be persuaded of the need to develop teaching that is disciplinary in focus, which then 'challenges EAP teachers to take a stance on how they view language and learning, and to examine their courses in light of this stance' (Hyland, 2016: 17), the disciplines themselves seem yet to be convinced of the need for a focus on language for academic purposes within their planned content curriculum (I suggest some reasons for this later in the chapter).

Here, then, I do not attempt to provide an in-depth analysis of the conceptual, discoursal or literacy expectations of any one discipline. Rather I hope to establish if and where the participants (both teachers and students) from across the three case studies felt that there might be a need for linguistic and literacy development alongside the teaching of disciplinary content and what they felt the focus of this teaching should be. In doing so, I hope to highlight to others where needs could be addressed that enable a way in to collaborative work between EAP and disciplinary practices.

Language and Content Knowledge

Despite Hyland's assertion that it is difficult to separate the teaching of content from the teaching of those skills which are a general focus of EAP practice and research (2016), those involved in EAP study and practice continue to attempt to do so. In the Routledge Handbook of English for Academic Purposes (Hyland & Shaw, 2016), there is a whole section on 'EAP and Language Skills', these 'skills' still being broken down, in broad terms into reading; writing; speaking; listening and vocabulary. This skills separation is also largely matched in the other section that has most relevance for most EAP practitioner contexts, with 'Pedagogic Genres' being broken down into sections on 'Lectures' 'Textbooks' 'Essay exams' and 'Seminars', for example. Thus, despite arguing for increased specificity, EAP literature continues to pay attention to generalised, skills-centred language work, leaving individual practitioners to interpret this for their own contexts. In terms of EAP testing and assessment, the difference between research claims around the importance of specificity and the reality of enactment becomes even starker. Flowerdew and Peacock state that 'Student assessment for EAP is the measuring of students' language ability' (2001b: 192), with no suggestion that this be placed or measured with consideration of a specific discourse or genre.

Content teachers, conversely, seem to pay little if any attention to language work in their discipline, beyond, as Turner suggests, an 'assumption that language gives immediate access to knowledge and must therefore be transparent' (2011: 29).

Within the context of UK Higher Education, then, we seem to be in a position where we cannot, yet feel that we must be able to, separate what we mean by 'language' from 'disciplinary content knowledge'. Rather than the nuanced and subtle understanding of language that Applied Linguists and those who research English for Academic Purposes have, the majority of those who teach in HE, including some EAP teachers, seem to take the view that language is

> concerned with standards, to assume and/or focus on idealised native English academic norms, and not to question whether these norms are the most appropriate globally or why they should still be considered in some way better. (Jenkins, 2013: 49)

This 'language' 'is often the catch-all term for problems with unmet standards, and the need for remediation' (Turner, 2004: 99). As a result of structural constraints, historical positioning and a lack of cross-disciplinary collaboration, it is this language remediation that is often delivered in an EAP classroom, entirely separated from other content learning that either will be or is taking place. This also seems to be what is expected from disciplinary academics (see email regarding Mai in Chapter 2). Furthermore, when asked to think about language as something to be learned, the majority will divide it into the fours skills of reading, writing, speaking and listening – as though these can be learned and developed in isolation of each other, and will also break these skills down into two key but very broad areas: vocabulary and grammar. In what follows, therefore, I consider my participants' understanding of where, how and if language fits into their disciplinary content knowledge, following their own focus on grammar, vocabulary, assessment practices and pedagogy (covering most of the four skills). In understanding these perceptions within a disciplinary context, it becomes more possible for EAP practitioners to foster collaboration.

Is grammar important?

While it is relatively easy for content teachers to create a list or glossary of lexical items that relate to the content of an academic discipline, and therefore meet some of the demands for EAP specificity, it is much more difficult, even for linguistic experts, to abstract areas of grammar away from content communication in a meaningful way.

Anecdotally, EAP teachers report that these discrete errors, such as spelling, punctuation and tense choice are highlighted by content tutors in the feedback they provide; Turner (2018) has also suggested that academics pay more attention to grammar than do writing instructors, but that they also underestimate the effort and time involved in getting it right. This focus on discrete items in feedback over more holistic discourse and argument development further feeds into both content teachers' and students' beliefs that there is an academic linguistic norm to be met, and that can be met by correction of and learning discrete items, relating to punctuation, spelling and grammar. This in turn meets a student and an institutional need for a quick and easy fix to language as seen as a problem. In this way, grammar becomes separated and disconnected from content and the sociocultural aspects of discourse building. EAP coursebooks often focus on discrete 'grammatical items' (what, in fact, counts as grammar as separate from other linguistic 'items'?) such as nominalisation, hedging or the use of personal pronouns, claiming to base their choices on corpus research that highlights this language as particularly academic. Lessons based around this language are easy to teach, easy to highlight in written texts when it is viewed as problematic and often receive satisfactory student feedback as it meets a perceived need for 'grammar' based on previous classroom experience. However, language taught in this way does not meet the specific needs of students nor develop an understanding of disciplinary choice.

An alternative approach to EAP teaching, taking content rather than linguistic structure as the basis for syllabus design can lead to an equally problematic situation, where teachers and students feel that there is no room for a focus on form(s) as language becomes subsumed under the content. Brinton and Holten report on a content-based EAP programme where dissatisfaction with the grammar and vocabulary component was a recurring and constant theme (2001: 243). Within this study, both teachers and students seemed to express a desire for there to be more focus on discrete item teaching and error correction, with clear guidance on how to do this and the specific items to be focused on provided within the course materials.

Grammar, then, can be viewed alternately as unimportant, surface level technical knowledge that should generally be hidden from view or brushed aside once the important content work has begun, or as the problem that needs to be fixed before the important content work can be undertaken properly. It is through perpetuating these perspectives of language work that Turner argues 'EAP has to a certain extent colluded in its own marginalisation' (2004: 96) and has maintained the false understanding around the (non)separation of language from content knowledge in academic study.

This difficulty in both separating and connecting language and content is evident in the transcripts from my project.

Feedback from students at the end of the summer pre-sessional suggested that some of them wanted greater focus on language, and specifically grammar:

- *It would be better if we could learn more about English grammar, which give nearly all Chinese students a headache.*
- *Maybe I need some classes to let me aware of my language mistakes.*

However, there is very little evidence from interviews that the mechanics of grammar were seen as a barrier to learning. When discussing this area of language, most participants struggled to articulate precisely what they meant or what they were looking for and shifted their focus quite quickly onto other skills. Thus, a teacher from the STEM site suggested that grammatical error was not really a key factor in writing success:

- *I suppose the UK students have grammatical errors as well ... and we're looking at things like their skills like referencing so contents of that is scientific contents and paragraphs and use of figures. (S1)*

'Home' students from both sites also touched on the mechanics of language. Here there is also a sense that 'language isn't important'.

- *it doesn't have to be very accurate but it makes sense and you know, answer the questions and it's just like an argument paper really, that's all that matters really. (C1)*
- *Because my writing's not that great to be honest, in terms of spelling and things like that. But yeah in terms of the style, very much I just try to base it on papers. I try to keep it very direct as well. Which is something that I've just learnt from science that you just like short sentence, like to the point. Partly that's imposed upon you by word counts as well because we're often bumping up against those. So yeah, I just try to keep it direct because in science you're just trying to communicate in the simplest way possible. Because a bit like what you've been discussing as well, science is a very international thing and most of its written in English and so you don't want it written in a strange way because you want everyone to be able to understand when English isn't their first language. (FS5)*

The need to write in simple sentences; the unabashed 'my writing's not that great … in terms of spelling and things' all come from an understanding that it is the ideas, or the research, that is important. However, this comes from a place of privilege; it is easy to dismiss something as unimportant when you do not find it a struggle and have never had to consider its worth. The message that academic value is not placed to any great extent on language use and accuracy was supported by the teachers, or at least by the messages that EAL students seemed to be picking up from comments made in class and assessment criteria:

- *recently I found that the lecturers just do not pay attention to your language because I was told that they just mark for your knowledge, your understanding and your referencing, I suppose, but they don't need to mark you for your language. So I was shocked. (5B)*

For students who came from pre-sessional classes or who had spent the time leading up to arrival in the United Kingdom focusing on meeting language proficiency requirements, this message took some adjusting to and also seemed slightly disingenuous given that, for the majority, it still seemed that language remained an obvious barrier to being able to access disciplinary knowledge. If language was not important, there was hope that the students' continuing sense that their English was not good enough would not prevent them from passing their assessments, but there was also confusion – if language was so unimportant, why did it still seem to be the main barrier to accessing the knowledge they needed? For some participants, this confusing message led to a stronger disconnection between language development and academic development:

- *although your language is not good but you still can use some simple sentence, the content, you have to do a lot of reading and to finish, yeah, so the target for us is to finish the essays ... So, we didn't pay too much attention to the language ... (6Z)*

The two main barriers to student learning from the student point of view, therefore, seemed to be around academic cultural practices and particularly the density of reading they are expected to do, rather than language as a separate area of knowledge. There remained, however, a contrast between the struggles described by students who were working in a language that was not their own and those from the United Kingdom, which focused very much on the difficulty of the content and the depth of thought involved:

- *after the first three weeks I was struggling in a way that the reading is very different, it's like up another level, it's very deep and philosophical at some point, and I couldn't understand it but it's fine. (C1)*

Within the STEM site, another 'home' student described the multi-modal nature of many of the papers they were expected to read. Here, she describes how it is possible to scan most texts at speed, and that what she needed to learn was how to 'read' the detail in the charts and diagrams:

- *And it's surprised me just how much information they can pack into some tiny little figures and results, you know. So normally if you don't need to read that in detail, you sort of maybe scan over it, have a quick look what it says in the conclusion but when you have to actually have to look in detail you realise that actually, you know, these tiny little pictures, the amount of work and information that's gone into it is really time consuming. (FS4)*

This level of detail and depth of thought is missing from students who were still working to read through dense text in an unfamiliar language:

- *the reading, has too many words, it's long and you have to pay attention to, you have to be patient and so there are some words that you don't understand and you have to check ... (6Y)*
- *Sometimes maybe not all the lectures that we can focus on. Sometimes I get distracted very easily but really it is very clear. Maybe my problem*

is not the understanding of the topic but reading the reading list and some of the books are very challenging and requires you more time. For example, one week we should read two chapters. Most of the time I can't manage to read all of it, maybe only half a chapter. And they build up. At the end of the semester I feel like I did not read as much as I have to. (6D)

- *I usually do the reading before the lecture. I find the lecture would be easier to understand if I do the reading first because I can know the key terms and their meanings from the reading. For each module, I often have two articles to read per week. I have three modules therefore I have six articles to read per week. I think it's as much as I hope to read. I don't have energy to read more and it's enough for me to understand the topic. But if I need to finish an essay, I will read more for the exact topic. Just as my tutor said, 'read as much as you can'. (Lin)*

Caroll (2015) has focused on reading as an issue for EAL students in particular, suggesting the reticence of teachers to name how many references are adequate for any given assignment can lead to cultural disadvantage. She suggests that for some students, 'many' references might equate to five texts which have been carefully read, understood in depth and learned, in comparison to the UK HE expectation of picking out key elements of a much wider variety of texts. Independent reading outside the teacher-fronted sessions of any UK TPG programme continues to play a key role in expectations around student learning. It is essential that all students have an understanding of the often-occluded practices that this reading involves; the linguistic understanding of concepts and grammar within the texts, of the linguistic choices the writers have made to express their thoughts and knowledge are an integral part of the understanding, interpretation and critical analysis of these texts. I would argue that it is here that grammar, and language more widely becomes vital. It is, as Turner (2004) has previously argued, and Caplan (2019) suggests more practically, an understanding of language *choice* that EAP teachers can work with students to develop. By considering the choices made by others, and the impact the choices they make themselves, a focus on form and accuracy becomes meaningful and specific.

Jargon and key concepts

For all participants, key terms, phrases or concepts were explicitly identified as problematic in terms of linguistic understanding. Students coming from a classroom-based language learning background, wanted lists of key vocabulary to learn, while many of the teachers focused on what they termed as 'disciplinary jargon'.

When asked about the language difficulties that might be encountered by EAL students in his discipline, the teacher who was new to the University responded:

- *I haven't thought about in much detail so I would say there is a lot of technology and jargon so things that I would use in conversation with the students so even things like DNA or protein which are really still jargon words or scientific words and I suppose I assume because they've got the knowledge whether UK or international so I would use those. With other maybe new terms or new techniques then I would try to, so the first time, I would try to find and expand stuff and basic principles and work from there. (S1)*

Although this participant uses the term 'jargon' here, as Thompson (2019) argues, this vocabulary is key to being accepted as part of a disciplinary community as it often has 'particular meanings' and allows 'insider' conversations to take place. Comparison across the Case Study sites reveals that there were differences in the way this jargon was understood between the two disciplinary sites, but also differences between the kind of language teaching that students are exposed to prior to beginning their studies.

It is clear from the response above that consideration of language was not high on the agenda for this STEM site. The teacher had not had any specific training or discussion with colleagues around working with international EAL students. Language was generally not seen as something necessary to focus on, but when asked to consider it, he was able to quickly identify vocabulary, concepts or terms that may be problematic.

Within this site, the focus was largely on key terms that are used to describe concrete processes or technological labels. These labels or terms were seen as linguistically problematic for a number of different reasons. These related to the complex nature of many of the technical terms, as well as the abbreviations that are frequently used to describe them. Thus, 'mas spectronomy' is shortened to 'maspec'; GFP is short hand for Green Fluorescent Protein. Again, these terms are loaded with implicit or presumed knowledge. They are also not easily pronounceable. Whilst Mai was able to recognise and understand the written terms, she was not able to make the connection with the written word and the same terms when spoken. This carries implications for the amount of knowledge she would be able to take from lectures and classes and, importantly, the implicit linguistic expectations placed upon her and other students in a similar position.

One example of the implicit, and often assumed knowledge within the STEM site came through two separate interviews. The first with the content teacher, who was considering his use of shorthand in technical terms as a potential linguistic barrier for international students:

- *sometimes there's something like gel or I might say a technique or something like masspec rather than using the full terms so with gel which is short for agarose gel, electro fereses or masspec is mass spectrometry and so maybe that's a bit like colloquial from when I was in a lab and speaking with people who were doing science as their job so maybe I should sometimes be careful about that. (S1)*

I questioned him about the meaning of some of these technical terms in general, and whether an understanding of them would be assumed or foundational knowledge:

- *we would expect them to have learned about it in theory at undergraduate level and so we actually get them to put their samples on using the mass spectrometer for this so yeah that would be a term that we would probably expect them to know. (S1)*

Yet, this term was also raised as an issue by one of the 'home' students:

- *we had exam on mass spectronomy, which I can't even pronounce, and I haven't, I know nothing or knew nothing about at all so I had to go and sit in the library and read up on that and find out about it because, but basically if I can get away with that, with not doing too much reading then there's always something more important to do, so ... It's a way of analysing, it's a very technical way of analysing proteins so they will put a protein into this incredibly sophisticated machine and it basically chips it up into little tiny fragments and then you have, you can analyse the fragments to identify the protein ... I don't need to use that technique but I need to be aware of it because I might be reading something where they, you know, they're saying, 'We've got our results from using this technique'. (FS4)*

Thus, within one (difficult to pronounce) word there is an entire process and understanding of a key technique which is assumed practical knowledge at the beginning of a TPG programme; it is not necessarily a technique that students will be required to use themselves but they are expected to understand it. The cognitive load involved in understanding all of this from the short-hand 'masspec' used in the practical laboratory session is large.

One further observed (and later discussed) issue for EAL speakers arose simply because the basic tools of the discipline are differently named, yet fully presumed knowledge. There is no universal name for the multiple types of microscope used for different purposes, for example. At TPG level there is a clear expectation that students have some knowledge of the use and different purposes of these different kinds of microscope, each with a different name. Although all students should, and do, hold this technical and conceptual knowledge it is unlikely that they will have been taught the different English names given to the different microscopes in their previous English classes. Within the IELTS testing system, although the word 'microscope' may be a vocabulary item that students taking the test might be expected to be able to recognise and recall, any further detail would move beyond the boundaries of a general academic language proficiency test. Therefore, in observation, it took Mai some time to understand the label given to a specific microscope within a practical laboratory session. A lack of understanding at such a basic disciplinary level is difficult to express to a large group, and Mai wasted significant amounts of time pretending to look for the correct piece of equipment, choosing the wrong one and then finally being given help to find and set up the correct equipment from the lab technician. The thinking and, to me observing, the clear feelings of shame involved in struggling with this one term then prevented Mai from listening and understanding the teaching that followed. Nor did the process she went through suggest to me that she had at least learned to recognise the terminology and use for the equipment she finally had, as there was very little conversation with the technician who supported her, and certainly no questioning or request to re-name or clarify its label. Thus, a lack of technical vocabulary led to further lack of understanding of the content knowledge that was expected to be built throughout the practical session.

A final difficulty was the discipline specific use of particular lexis that would be taught in language classes as having a more generic, slightly different meaning (see Hyland & Tse (2007) for further consideration of this issue). Thus, terms such as 'architecture' and 'expression' are used in biology to refer to cells and genes. If taught at all in an English language, or even general EAP classroom setting these words would be very likely to be given a differently oriented definition. An EAL student, therefore, is required to transfer her own biological content knowledge and apply it to a different understanding of an English word that she may or may not have already encountered and then make the connection between the two. The cognitive load required to understand something that is presumed knowledge in the discipline can, at times, be so high that there is little energy left to build on this knowledge and think beyond the terminology.

Although problematic, the majority of these content related terms are arguably relatively easy for students to learn, presuming they already have the disciplinary knowledge required to understand them.

A biological dictionary or tab on the VLE supplying commonly used terms and their meanings may go some way to enabling EAL speakers to move beyond this linguistic barrier to learning. This would be further aided if used in conjunction with teachers having some awareness of the difficulties these labels may carry and taking them into consideration in their teaching. Elsewhere in the University, for example, I have seen labs where all the equipment is labelled, and laminated glossaries for key equipment are placed on each bench, enabling students to listen, look and identify basic equipment with ease.

Thus, within this site it appeared that little consideration was given to the place of 'language' within disciplinary knowledge building and understanding. When asked to consider how it might have an influence, responses from teachers were generally centred around key terms and technical labels. Conversely, in the AHC site, it was clear that teachers had already given the place of language in their discipline significant amounts of thought. Many of them had an acute awareness of the power of language in terms of their own disciplinary research, with, for example, Critical Discourse Analysis being one of the methodological approaches they both taught and used. Thus, when asked to consider where they felt language knowledge and content knowledge intersect and interact, responses were more nuanced.

What emerged from the responses of teachers from this site was, to some extent, an idea of what they understood a language teacher (i.e. the researcher they were speaking to) might be able to do to help them support their students. To this end, they seemed to attempt to separate out discrete language items that they felt might be teachable in a 'language classroom' from the more discursive or communicative difficulties that they later went on to consider. Thus, although the teaching participants from this site had quite a sophisticated understanding of how language influenced access to, understanding of and communication in their discipline, when they were asked to focus on what this language might be, they also chose to focus on key terms.

In this discipline, the key terms *were* more abstract and complex theories or concepts rather than technical processes or equipment, so were not so easily definable, and therefore not (in theory) so simple to teach.

- *Everything has jargon. So, in the academy we have jargon. It's academic jargon. Each discipline or field will have jargon of its own as well ... So that's completely alien to many students, and so they're asking what these even individual words let alone phrases and concepts actually mean ... Some even don't know what normative theory is, which would be the same across disciplines in the UK. Explaining that to someone who finds it hard to understand language is very difficult for us as academics, and hugely time consuming. (M6)*

Here, understanding is beyond the linguistically translatable or definable. One difference between students who still view themselves as low-level language learners and those who feel confident with a language is in their confidence around how to distinguish between words and phrases that are easily definable and less complex and those that require thought beyond a dictionary definition, into an exploration of the theories of their discipline. Once you are comfortable and confident in a language, you are more able to tolerate ambiguity and accept that developing an understanding around the 'particular meanings' of conceptual terms is part of your disciplinary knowledge construction.

- *Throughout this semester I encountered with many, no, a lot, many alien concepts and theories, but sometimes you just have to accept that you don't understand. (C1)*

However, it was not simply theoretical and conceptual language that provided barriers. Students described how cultural references, loaded with hidden meaning and understanding also added to the time they needed to spend on reading and interpreting a text:

- *So when you just read a paragraph you will find many evidence of it. Like if they say 'the Cold War', I said I should read it. I know it's common history knowledge but sometimes you don't exactly know what happened there and how this affects this idea in this paragraph. It is understandable but it takes too long. (5B)*

At TPG level, the issue of key terminology is often compounded by students entering a programme without having any previous background in the subject. As AHC programmes can draw on a more eclectic prior knowledge base, students from this site often move across disciplines entirely from their undergraduate to their post-graduate programmes. Thus the 'normative theory' referred to by M6 above will be both new vocabulary and new knowledge, yet will be assumed knowledge for students embarking on a Masters programme in this discipline.

Even when academic qualifications are scrutinised and appear to closely fit the required profile for Masters level study, this is not always reflected in reality. Thus, Mai's personal tutor claimed that she had scrutinised Mai's previous qualifications and concluded that she would begin her TPG programme with the required subject-specific knowledge, yet Mai disagreed arguing that her undergraduate degree had provided her with different knowledge that her tutors were not aware of or taking into account.

The necessity of using key words and understanding how to search with these is highlighted in both disciplinary sites, as is the need to scan through a large volume of papers in order to find relevant and up-to-date information. Students from the UK, who had also studied for their undergraduate degree in the UK, attempted to explain how they used their understanding of key subject vocabulary to search for information:

- *the first thing you do is if you got a new topic is probably go on Wikipedia to be honest and see what's there. Then at that point I'd use Google Scholar and look for some good recent reviews. And then I'll just do things like just use loads of different words related to it and I'll probably open like 40 or 50 different papers just from tabs, just do search. And then I'll look at the papers that have cited those papers, always try and bring it up to you know as recent as possible, 2017. And then I just get my note pad out and then go through all these papers and write the title down, write the useful stuff from them and I'll do that for like 40 papers and then through that you identify like the best ones and you've got all the information laid out and then at that point I'd go on to write it. (FS5)*

and then how this fed into their developing understanding of how to communicate this themselves:

- *... through reading the papers. So, you try to write like they've written them. And then just through an extent through trial and error through undergrad because you've got first year to make all these mistakes and they're like 'No, you don't write like that when you're doing the science thing'. (FS5)*

This contrasts with Lin's reported experience where she felt that not having any of the necessary key terms available to her from previous study also prevented her from further development:

- *the understanding of language sometimes have a negative influence when I try to search some materials. Because I do not know what the key word is in English. (Lin)*

Furthermore, even when students felt that they had the disciplinary understanding, their lack of confidence in their language fluency prevented them from fully demonstrating this to others. Here, a student from the United Kingdom describes the complex process involved in working together to build an understanding:

- *there are certain terms or some words that sometimes I don't get and they're able to help me with it. Obviously when they're trying to explain themselves I have to really try and listen and understand because I'm almost giving them the words when they're trying to explain something to me. (C2)*

For those students who were operating through English as an additional language, the complexity of the process around expressing disciplinary understanding through English is clear:

- *the speed and sometimes maybe some academic words, vocabulary, and also maybe – so it's complex, mixed ... (6D)*
- *I think it's in many, many places, the grammar, the word and my thoughts, I think is not very ... I don't know how to say in English ... Critical ... (6X)*

Here, the student identifies the transference of thought across languages as resulting in a loss of criticality; the nuance of meaning and depth of understanding is lost in a vacuum of words. A student who had previously studied through English as a medium of instruction in India had a similar experience in the STEM site:

- *I think it's because of the vocabulary because even if we study in English all the instructions are in English but we don't usually use English, we just speak in our own dialect so I find it difficult to use different kinds of vocabulary to make it. I mean I just use only simple words. (FS1)*

The focus on problematic disciplinary vocabulary used to name key concepts, processes, techniques and equipment in the target language (English) as being the main 'language' difficulty is hardly surprising. This language is easily identifiable and 'knowing' and feeling at ease when using these terms is clearly a key aspect of gaining access to target knowledge.

This ease though, needs to transfer in two directions. These students are developing multilingual disciplinary skills, and the majority of them will return to their home countries for future employment. They therefore would also expect to be comfortable communicating their knowledge and skills in both languages. For some students, a fear of losing their home identity and language could go some way to explaining possible resistance to learning in English. A number of students reported similar experiences to that of Mai:

- *sometimes if you use English you can't remember the Chinese name.*

Very often, in fact, students reported having no knowledge of the vocabulary they were learning in English in their own language. They therefore had no conceptual background on which to hang the language they were encountering.

A simple response to many of the issues discussed here is for teachers to provide lists of key concepts, with simplistic definitions.

Using a shared space for this would allow students to add and develop this glossary in partnership with each other and their teacher, adding depth and richness to the definition. While this may help all students to develop a basic understanding of disciplinary theories, true understanding should develop through seminars, lectures, practical work and reading. Guiding students through this process of understanding should be part of the pedagogy of the discipline. Corpus researchers would suggest that knowledge of how to use and manipulate a corpus of disciplinary language would provide EAL students with the tools needed to develop a deeper understanding of disciplinary terms with relative autonomy. Others might suggest that there simply needs to be greater tolerance for the use of *any* language, allowing students to bring their own language into the learning environment and engage in code switching without feeling a sense of shame that they are not speaking in English at all times. Turner (2018: 14) has suggested that there needs to be an institutional wide ideological shift away from a 'linguistic conflation ideology' where 'problems with linguistic accuracy, as well as style, are seen to index intellectual deficit'. Language, therefore, needs to be considered as part of the academic learning process.

In fact, without a continued, supported focus on language development (both lexis, grammar and skills use) throughout a TPG programme, many EAL students reported a *decrease* in language proficiency and confidence over their year of academic study. English is used to communicate disciplinary understanding; if the test of this is only via written assignments, then students focus almost wholly on language use in this context; therefore:

- *I think the language I get improved in my writing and in my reading but I think my speaking is getting bad … Because it's not like the pre-sessional that you will have a chance to discuss in the Reading Circles, [Seburn, 2011] for example. So, most of the time I'm either just listening or writing. So, I get also worse in my own language, I think, because when I – I think I don't speak very much the whole week, only weekends, maybe. So, I don't know. I think I was more confident in my speaking than before. And I'm thinking about what I'm going to say. It's not like I'm slow and things. (6Z)*

When students find themselves a part of a majority minority, as do many Chinese students studying in the United Kingdom, the impact on their language and academic development could be profound:

- *Most of the class mates are Chinese. Sometimes, I just feel like I study. Last week, in a lecture, I just saw all Chinese there and I forgot to speak English and I was just speaking Chinese. (5A)*

This same student reported an overall lack of improvement in her English language:

> *5A: But I still think that the grammar is really terrible. When I wrote my essay I just feel really confused. And also my vocabulary. And I don't think that my English has improved.*
>
> *BB: Do you use English outside the classroom at all?*
>
> *5A: Seldom.*
>
> *BB: So, once you leave and you have an hour – so you have an hour for your core module seminar and an hour for your optional module seminar and then lectures and outside of those you do not use English except in shops.*
>
> *5A: Just to write essays.*

Therefore, the student participants and, to a lesser extent, the teachers, all found it difficult to express how language and disciplinary content interacted with each other, or where it was the content knowledge and where it was language that caused problems in accessing the curriculum. For those who were already expert users of the language and had not previously worked with students who were not, language remained under considered. For those who were still developing an expertise in English and for those who had worked with large numbers of EAL students on their programmes, language was viewed as an essential issue:

- *The extent to which language is a, language competence is a necessary tool of doing, of completing the programme is very high. It's absolutely essential because the programme's focus is bringing people to a stage of being able to expertly critique and debate the issues that we're discussing and that we discuss them in English and read about them in English and engage with the experts in English and therefore the ability to personally reflect that through conversation as well as writing is absolutely essential. (M5)*

The questions that remained were over the nature of the best approach (or even whether to approach) the teaching and learning of language within a disciplinary TPG programme. Whilst seen as intertwined with knowledge communication, academic teachers across both sites still maintained the ability to separate out language from content. This assertion was particularly strong when they were asked to consider the role of language in assessment practices.

Language and Assessment Practices

Whatever format an assessment takes, it is difficult to conceive of one that does not involve a requirement to interpret and communicate knowledge through the use of language at some stage of the process. As formal assessments are the only means by which a student is able to prove they are worthy of being conferred with a Masters degree, it therefore seems logical to assume that language competence, and specifically the ability to manage disciplinary discourse, is a necessary element of any TPG assessment process.

While there is now a growing interest in assessment in Higher Education, little attention seems to have been paid to the place of language and whether it should be a focus of programme or module learning outcomes, and therefore assessed. When language is considered, the focus is either on ensuring clarity of instruction in assessment tasks (so on the teachers' use of language) or suggesting that it should not be considered as part of the assessment rubric as the intention of the assessment process is to measure disciplinary knowledge and understanding (therefore perpetuating the myth of language as the transparent carrier of content).

In this section I provide a brief overview of the current assessment landscape in the UK HE context, attempting to position language assessment within this. I then draw out both teacher and student perceptions around assessment from my data in order to add to this picture.

Assessment in HE and in language teaching

'Assessment is broken in higher education!'

Phil Race (2015) suggests six factors that have contributed to him drawing this conclusion. Two of these relate to an improved understanding of the purpose (learning) and validity of assessments, one is connected to technological developments and the other three relate to an increase and change in the student population. Here Race makes specific mention of a more global student body: 'The world has opened up, so that our assessment processes and practices need to be more compatible with those in markedly different cultures and traditions'.

It is often this global student body who are blamed for the fact that assessment in HE is broken. Newspaper headlines around the increasing numbers of academic integrity cases frequently highlight a growing international student body as the cause of the problem, conflating 'international student' with 'language learner' again (see for example Bretag (2017) and the Times Higher Education's write up of the same research in McKie (2019)). Although many may argue that this fear of international students causing a crisis in HE assessment is misplaced, it cannot and should not be dismissed. In their literature review of research into inclusive learning and teaching practices more broadly, Lawrie *et al.* (2017: 8) found a 'sustained theme that emerged ... is the extent to which stakeholders, particularly faculty, worry that inclusive assessment practices may reduce academic standards and erode educational quality'. It is therefore necessary to address these worries directly in order to be able to move on from them. In fact, it is arguable that these worries are driving the increased and overdue interest in establishing assessment practices that are valid, fair and transparent.

The EAP and broader language education community has, in contrast, a longstanding concern with both testing and assessment. This is in part because EAP teachers, and the Centres in which they work, have often traditionally (and incorrectly) viewed themselves as linguistic gatekeepers to the academy. Many of the current debates in Higher Education are reflected here, with practitioners being urged to develop their 'assessment literacy' (Fulcher, 2012). While EAP teachers may have a more developed understanding of issues relating to (language) testing and assessment than the majority of our counterparts in Higher Education, this does not necessarily result in a greater confidence in our ability to develop good assessments ourselves. Much of the literature on language testing and assessment, whilst developing the ability to critique tests and analyse them for their validity and reliability, has much less focus on how large-scale testing processes can become smaller-scale, contextualised assessments:

> Even when textbooks discuss the needs of classroom teachers, they frequently describe techniques that are drawn from large-scale standardized testing, many of which are not applicable to the classroom. (Fulcher, 2012: 116)

Moreover, as EAP practice uncomfortably straddles private, profit making and globalised agendas and the more optimistic aims of providing an internationalised and socially just education for all, assessment literacy has often focused on preparing students for external tests such as IELTS rather than developing an expertise in assessment within the context students are going to enter. Attempts have been made to develop agreed measures of proficiency across all languages, the most successful being the Common European Framework (CEFR), which suggests measuring learners of any language within a specific domain (for example professional or educational), across different activities (interaction, production, reproduction and mediation) and considering linguistic, sociolinguistic and pragmatic competencies. For EAP teachers working within HE, it seems to have become necessary to translate any student assessments across three different measures – those of the academy; those of the (usually IELTS-based) language proficiency entry requirements and those of the CEFR. Assessment literacy has become more about understanding and interpreting these measures than developing a theoretical and practical understanding of how to develop assessments for our students that act as a good driver for learning (Race, 2015).

Furthermore, the focus of EAP assessments remains specifically on measuring language proficiency (Flowerdew & Peacock, 2001b) and is concerned with entry level requirements (Murray, 2016a) more than the continued and ongoing assessment (and therefore learning and development) of students' language and communicative competence as they work through their academic programme.

In this way, the binary and artificial separation between assessment of language at point of entry and then of disciplinary knowledge with no acknowledgment of the place of language within this communication process, perpetuates the separation of EAL students, and others who do not find academic linguistic expression easy to manipulate, from the 'rest' of the student body.

The teachers' perspective

My University's Code of Practice on Assessment suggests that 'assessment criteria will make clear if, how and where accuracy in written expression and presentation are taken in account in marking and this will relate to stated learning outcomes' (University of Leeds, 2017). On most TPG programmes, assessment criteria worked to assess 'presentation' or 'written expression' (within which there was an implicit but very occluded focus on language) as a separate measure to, for example, 'intellectual skills' and 'knowledge and understanding', although it was unclear where the former were taught explicitly. This required those involved in assessing student work to have the ability to separate both linguistic error and linguistic elegance from the communication of content knowledge and disciplinary understanding. The question of how teachers approached this conundrum is arguably key to understanding where perceptions lie in terms of the connection between language and disciplinary knowledge.

All participants claimed that they were able to make this separation, yet few were able to articulate how they had learned to do so other than through years of practice:

- *Marking for content only – it's sometimes hard to assess their level; how to do it evolves with practice. It's a shock when you first read some of the essays, you need to adjust – not lower your standards – but learn to see the argument which is there but not always expressed very elegantly. (M1)*
- *I remember the first time I was an external examiner and only just started post graduate teaching here, so I hadn't really marked any essays, and I saw some of the essays from the University of XXX, who also have an enormous international TPG cohort. I just thought, 'How*

on earth can they mark any of this? It doesn't make any sense'. I was really shocked. Then 18 months down the track and you don't even blink. (M10)

Less experienced members of staff were less confident in their approach, but maintained the same argument that experience and intuition, alongside some guidance from colleagues had enabled them to see where a student was experiencing difficulties or showing understanding in either language or content:

- *I think I can most of the time, but I often feel inadequate in doing so. It's not at all uncommon, in fact I would say it's frequent that we have students who are struggling on both accounts and I think it takes a degree of experience to be able to pick apart where the major problem is. (M5)*
- *I don't know if I can articulate it. I've taken that message on board and that is what I attempt to do; to read the content and discern comprehension. Yeah, the easy answer is to say I just read it and, due to my years of experience, I can intuit whether they've understood it, even though their language is dreadful. That's how it feels, I suppose, as I'm doing it. That's not a very professional answer. (M9)*
- *we mark on clarity, that is part of the assessment so that is obviously taken into account when we're marking, it is very difficult to do because of the large numbers of international students that we have but generally I found it quite difficult when I first started but I'm used to it now I feel and you do tend to mark the content. (M2)*

Here though there was a clear sense of discomfort around not being able to fully articulate the process they go through, and around the multiple difficulties and imprecisions it involves. It also seems that, although the assessment criteria did not explicitly cover linguistic ability, and although all agree that they had developed an ability to see beyond poor language, it remained impossible to fully separate the two. Clarity, written expression and argument were all assessed; all of these involve linguistic expression; all of these are the 'polished prose', the 'taken-for-granted "good" things' that are given no credit for being there but suggest intellectual deficit when the language draws attention to itself rather than the content' (Turner, 2018: 24).

- *So they're being marked for content but even the task of expressing a complex idea at that level, you know, they have to still convince the marker that they've understood what's going on and that's tough. (M4)*
- *I suppose it's about clarity, isn't it? Equally, it would be very unusual to get the top grade and have not so great language. I don't think that's really possible. But I don't think a particular value needs to be attached to that. (M10)*

- *it basically gets to a point where you're just like reading something and going, 'I don't know if I've got any idea what that person's really trying to say. I really just don't know', and that's the point at which we as a School feel that we can't award marks. If we can work out what we think they're saying without making what we think of as too many assumptions on the way then marks will be awarded accordingly.* (M4)

A number of teachers described going through a shift in their approach to marking, moving away from a focus on correcting the easy to spot mechanical linguistic error towards a realisation that structure was more important:

- *When I started, I'd be correcting every aspect of grammar and I don't do that anymore ...* (M9)
- *It's probably a bee in my bonnet but it's to do with fully constructed sentences or straight clauses and that kind of stuff; more structural stuff, I suppose, than worrying about apostrophes and commas and paragraphing structuring as well – Make sure you've got one key idea per paragraph – it's that kind of structural level grammar in the very broadest sense, I suppose, is what I will tend to comment on, aside from the content side.* (M9)

Others used their years of experience as a reason to dissent from the suggestion that the language and content should and could be separated, whilst maintaining that they could provide a fair assessment within the current assessment criteria:

- *I am personally troubled by the focus upon content as opposed to the broader meaning that's being conveyed by students. I find it somewhat unhelpful but I do think that our criteria we have is sufficiently flexible to allow us to take into account both the students' ability to communicate clearly as well as the indication of their, of specific bits of knowledge. I have found that over, well, probably 20 years now of working extensively with Masters students who have English as their second language, second or third language that it's unhelpful if not impossible to judge the simple absorption of facts without also taking into account the ability to express and analyse and critique those facts.* (M5)

There was agreement that language and the ability to present a convincing disciplinary argument were frequently so closely connected that if a student was unable to manipulate the language effectively, the argument was likely to collapse as well:

- *an awful lot of papers that either fail or just barely pass where there's a great deal of that kind of sort of mechanical error in English usage but it's very often largely combined with incorrect word choices and often incomprehensible logic within sentences. The structure is so convoluted that almost every sentence can be read in multiple ways. (M5)*

Furthermore, teachers were also acutely aware that when working with diverse cohorts of students it was important to be vigilant in terms of your own biases and assumptions around language use:

- *You can get students who are very, very able speakers whose written style of English is actually quite poor. If you're marking anonymously, you can get quite a shock and think, 'Oh actually, I thought that student would do better than that'. (M8)*
- *I don't tend to mark at all for language because I just think that you can also end up with the flipside which is a student who is linguistically very competent but who hasn't got the ideas. (M8)*
- *I think that we have this obsession with language, which is terrible ... I have seen essays which are beautifully written in style but absolutely say nothing. And I've seen Chinese students producing the most incisive critical interesting piece that could be published anyway but gets a bad grade because the language was not there completely ... but I also think we need to make sure that when we assess, we are able to assess substance. (M3)*

Within this, though, there remain dangers of making assumptions based on typical language error transference from a first language. Institutional policy on anonymous marking is intended to remove this kind of assumption and implicit bias, yet if teachers report reading an assignment and making assumptions around the first language and nationality of a student, they are likely to also be making assumptions around their academic abilities, particularly given that so few teachers seem to be able to fully separate one from the other:

- *Well certainly because we get the large percentage of the international, of the second language students we have are Chinese there are a number of common characteristics in terms of the kinds of errors that are made, which are often about missing articles and prepositions and poor sentence structure. When that exists just to a fairly limited extent it's quite obvious that it's a Chinese student, I'm marking an anonymous paper, by seeing those kind of things but the, most meaning comes through in most sentences then it's still possible for the student to have a fairly strong paper if there's also evidence of some good understanding in there and I see a fair bit of that. (M5)*

Ultimately, the approach to reading and assessing work in this site is summarised by two distinct voices below:

- *And clearly we shouldn't be giving people MA's for writing nonsense. But I mean, you know, I guess in the language of linguistics, I am more interested in pragmatic expression than in syntactical expression, and semantic expression. (M7)*
- *there is a Chinese way of writing English, that's cool. We're okay with that and we tell students that. (M4)*

Despite the claim that *'we're okay with that'*, there was a real sense of frustration around the difficulties of assessing international (increasingly focused on Chinese) students on the scale involved. This frustration, again, was based not only on the quality of the work they were grading, but also on the time it took to do so.

- *they often go to quite obscure things; we can't take anything for granted so we will go and read those sources and that's why it takes a long time. But when you go to try and read the source and it's in Mandarin you're just, 'Well I'm sorry, I can't do that'. (M4)*
- *The kind of papers we've been talking about take an awful lot longer to mark than, you know, than even average quality papers written in...... reasonably good English. (M5)*

The pressure to gain good grades without the developed understanding of a subject and a perceived inability to fully engage with the content were seen to lead students into difficult situations:

- *it should be said it also goes frequently beyond that to the point where students, this is speculative but probably because of the perception of pressure on them, will attempt to construct content through artificial means that ends up getting them into a lot more trouble than they would have otherwise been in if they'd been original. So, I'm not necessarily talking exclusively about plagiarism, although there is a lot of that, but I am talking about using translation software or sometimes doing that in combination with a lack of originality and just coming up with a lot of largely unoriginal and very often nonsensical writing. (M5)*

These time concerns extended beyond the assessment process itself and into the administration involved in arranging resits for those who did not pass at first attempt:

- *So it's not, you know, it's not like, 'Oh you failed. Bye, see you later'. No, we're not doing that. We're saying, 'Oh you failed so now let's*

> *help you resit your dissertation so that we can get you through'. And if you look at the feedback that students are getting] on those failed dissertations ... you will see that they are getting explicit, lengthy, written feedback, you know, in order to pass this dissertation. (M4)*

The implication being here that teachers work hard to support students to pass first time around in part because the impact of multiple TPG failures on the academic cycle of work would be devastating, with a new cohort already in place whilst resits were taking place, and summer research time rapidly vanishing.

Again, growing numbers of students, and particularly growing numbers of students who were not immediately able to access the learning culture of a School were seen to be having a radical impact on the perception of TPG education. Although teachers reported feelings of discomfort in class-based teaching situations and a need to question their pedagogical approaches with this changing cohort, it was within the assessment process that difficulties were most brought to light. Here is was impossible for an individual student to hide, and teachers had to take note of both the language and the discursive abilities of each student. It was here that they were forced to notice that maybe their teaching had not been effective, and that language was having an impact on how well messages were conveyed.

The students' perspective

Much of the assessment in Site 2 (AHC) was in the form of a traditional discursive essay, and this increasingly preoccupied the student–participants throughout the lifetime of the project, as they encountered disappointment or surprised relief in the grades they were awarded.

Concern around writing was, in fact, more connected to reading and understanding how to incorporate the work of others into their own writing. It is clear that the students paid real attention to the feedback they received on their first pieces of writing and found this informative and developmental to their understanding of academic requirements. One student in particular was greatly disappointed with the grade for her first essay. The feedback she received focused on a lack of cohesion in her writing, and we discussed at some length what this meant and why this might have been the case. She concluded that, when writing, she had felt that her ideas were clear and cohesive, but that her way of thinking was different and she needed to understand how to write differently; again, her own conclusion was that this was due to her previous educational culture, not language difficulties. She clearly took the feedback very seriously, and was determined to develop her ability to write an acceptably cohesive argument:

> ***5B:*** *I didn't feel like I've perfected this skill yet. I feel like, for cohesion, I don't scatter my work but instead put some files and organise them into smaller topics but all categorised with every argument. It makes my brain also organised, so when I'm done I can follow where every argument is.*
>
> ***BB:*** *How did you learn how to do that?*
>
> ***5B:*** *From my friend, actually. She's in the Business School. I felt like my way is good but it takes a long time and it's in one document file listing everything. She showed me her way of doing things and I like it. I tried it.*

Students encountered difficulties when writing was given different terminology and felt that they could not transfer discourse skills learned from one assessment genre to another. Thus, a literature-based essay covering largely theoretical arguments was viewed as having little in common with a Case Study, therefore requiring the development of new skills:

> ***5C:*** *I'm struggling with the case study because at the final essay … I just passed the essay. XXX gave a lot of advice and pointed out lots of problems from my essay. I had no idea about how to research a case [laughter].*
>
> ***BB:*** *Is it because it's a case study and not an essay that you're not sure what the difference is between them?*
>
> ***5C:*** *Yeah. I'm good at the theoretical analysis but not the case study.*

There were differing approaches to the writing process in general, and the journey of understanding this took them through. This ranged from a formulaic, 'paint by numbers', instrumental approach in order to get an assignment completed:

- *It's mostly presented like cooking. I usually write my essay outline first and now I think if I can plan what I want to do, like cooking and at every stage, I'm sure what I want to do and I can cook well. (5D)*

In contrast to an understanding that the purpose of an assessment was to encourage further thought and learning:

- *the assessment that they gave us is one at a philosophical level, so not really practical so it's like, you would be expected to read and then kind of … but there's no real right or wrong and the question is there just to guide you. (C1)*

- *In order to finish this essay, I did a lot of reading, but I have to say I have more questions than before. I think what I thought before is too simplified. (Lin)*

For many students, their assessments created a real sense of a developing understanding. Students were aware that they were learning, but also that it was a difficult and often confusing process. However, it was still not clear to what extent this confusion was connected or not to their studies being undertaken in a second language.

In contrast, student participants in the STEM site had a highly instrumental and practical approach to their assessments. Time and timing of assessments was a concern for all, as was clarity of instructions and the use of assessment criteria.

Here, students felt that those from the United Kingdom had an advantage over international students in terms of educational culture and knowledge of how to 'play the game':

- *it's so different here because even the marking scheme when I come here I thought I have done my best but I got, the score I got was much less than I expected, so I feel like my level of writing is so different because when I see my classmates who were from England they are much, much better than me. (FS1)*
- *one of the things that they always said at the start and which, obviously coming through the English education system you know well, is like look at the mark scheme. And we're 99 percent of the time, we're always on the same mark scheme. And at Masters level 40 percent is the content and 40 percent is critical analysis and 20 percent is like presentation, references. (FS5)*
- *if you're submitting something you want it to have 2017 on there. Just from the mark scheme, it doesn't even matter if it's that relevant or not. You're just getting in there … So, it's not a great way of doing things, but you play to the mark the scheme so. (FS5)*

UK students were aware of expectations and had experience of working towards assessment criteria which international students did not. This was also recognised by one of the teachers in interview:

- *we sometimes have tutorials with the students as part of another module on scientific communication and when they submit an essay you could see differences … things such as referencing, how they write, even things like putting paragraphs and the use of figures and maybe even just their academic style. We'd tend to write in the third person rather than 'I did this' and, yeah, I don't know whether that is a difference between the UK and inter but possibly yeah maybe just based on the groups that I've taught maybe students that have been to ... or another UK institution tend to have similar referencing rules and writing whereas that might be different to maybe international universities. (S1)*

In this site, though, it was not only the unwritten, implicit expectations around disciplinary communication that lacked clarity. The purpose and instruction of some of the assessments were also questioned.

- *I did XXX as one of my other units and we were set an assignment I didn't even realise we were set 'cause it was on the VLE somewhere and we hadn't had an introductory lecture or anything. (FS2)*
- *some assessments that I feel like have just been a little bit lazy on the part of university and I would say that kind of is like, it feels a little bit lazy when we put a lot of work in that some of the assessments are just a bit like not that well thought through some of them. But I guess they do it to pad it out a little bit, but that is what it feels like with some of them. (FS5)*

For these students, this lack of clarity and purpose was frustrating and time consuming but had little other impact. For students like Mai, who need to spend much more time reading and working out the value and meaning of each task set, to be given 'lazy' assessments that are not fully thought through could have a profound impact on their ability to complete a programme successfully.

There were, though, also some instances of good practice being highlighted. This was almost always in relation to clarification around where marks will be awarded:

- *for several pieces of work we've had an actual tutorial schedule, the purpose of which has been to explain what we had to do. So most of it's been really good actually. (FS4)*
- *Usually we've had somebody explaining in detail so ... We've been given a breakdown of where the marks are being awarded for each section and how big they're expecting each section and what, you know, they've given us a lot of advice about what to do. (FS4)*

However, this explanation was given orally, within a group context, with the expectation that all students were present and that they were able to follow, understand and ask questions when needed. This is not something that all students feel able to do regardless of nationality, but more typically if they are an EAL student who is working through complex and unfamiliar language as well as ideas. Having observed one of these sessions, it was clear that a few students, all of whom were first-language English speakers, dominated discussion and question time, with no direct or individualised comprehension checking of those students who did not speak up.

Comments around feedback were also mixed and contrasting. FS1, an international student, describes using the feedback from her first assignment to develop her understanding of the academic expectations and requirements. Thus, despite her initial disappointment at her grade, she was able to use the feedback as part of her learning process:

- *I think they really give good feedback because when I got the mark, when I saw my score I was like oh, I expected better but when I read the feedback I felt like okay I deserve this. (FS1)*

Other students were less complementary. All agreed that they received feedback, and that it was taken seriously by the School, but they questioned its use and the depth of care that went into it:

- *I read the feedback. And it's variable, some that I would say is not that impressive. Sometimes I've read the feedback and like they've not been reading it properly. They've not been reading the work. I mean the grant proposal, there was one bit where he was like 'You never mentioned this before?' and it was like two lines up. And it was like literally it was right there. I feel like they're not reading it sometimes. Because I think they're marking a lot. So how useful is the feedback? ... Like I always read it because I find it very useful. But I can't say that there is any big thing that I've taken from the feedback here. (FS5)*

There was a sense that initial good intentions from their teachers were abandoned as workload increased throughout the year, or that the feedback provided was not useful for future development:

- *That only happened the once and then the work we've had back since, it tends to have some ticks through the work where presumably they've found something that they agree with or they like and a couple of little comments but not very much in-depth analysis really. (FS4)*

Again, the sense of a lack of engagement in broader student education practices seemed apparent. The student below suggested not only that the feedback provided was in order to tick required boxes, but that there was little encouragement for students to make use of 'Office hours' in order to gain further clarification or support

- *I think feedback is emphasised here as important but although we do get feedback on every piece of work, as a rule or like a regulation or something maybe just … One of them was literally, 'Results do not match conclusion', and I was like, 'Oh, okay', but she didn't tell me what the real, what I should have written. It was just that was wrong and then I can't apply that for any other situations really and I don't know if I can go and speak to her about it. I don't know, she probably, she has office hours that are probably good but I wouldn't probably be there. (FS4)*

For students who have the academic confidence to ask for help, to find support amongst peers and to read instructions and make linguistic guesses to enhance their understanding, these practices may be frustrating but surmountable difficulties. For international students, operating with the minimum level of language proficiency and within a new academic paradigm, it is vital to ensure as much clarity as possible around assessment and feedback.

Whilst some of the academic expectations at TPG level were also new for students from the United Kingdom:

- *we had some lectures at the beginning of the course that were about scientific communication, about how you're expected to write various different things and they were talking about criticism there. So yeah, it's something I had to sort of think about, how to write the, with the, on that basis. That was something new for me as well actually, yeah, but it's something that I've been okay with. (FS4)*

International students faced difficulties that extended beyond the development of their own critical voice, as they learned to communicate across the multiple genres required by their assessments:

- *when I come here we don't have many lectures but we have a lot of self-study and do a lot of critical thinking which I'm not used to because I've been used to examination system so just before the examination I just have to read a lot, learn it by heart and go and write it so I don't have to think about it but here we have a lot of assignments to put in our ideas and that's a difficulty. (FS1)*
- *for assignments they just give us a topic and we were asked to write a literature about it, so for me I had no idea what to do so it was just my first time, so I Google what is literature, what to do about it, what are the contents that I should write and I think it will be better if … I know that most of them know what literature is, most of them have done it so they know what would be the contents but since I don't know it would be nice if they have to like little bit about what it is, explain about the assignment and what they expect because since I don't know what they expect I don't know what I write, it was just too difficult for me in the beginning. (FS1)*

Again, these issues appeared not to be connected to linguistic proficiency in isolation, but clarity of expectation and cultural understanding, conventions that are treated as 'common sense' and communicated through wording as if these were 'transparently meaningful' (Lillis & Turner, 2001: 58). One UK student described helping a friend from Indonesia at the beginning of the programme who had misunderstood the advice provided on avoiding plagiarism:

- *she did find some of the written work difficult. She particularly, we had some talks about plagiarism and advice and stuff about plagiarism at the beginning of the term and it was explained, you know, 'You can use a direct quote but you have to say it's a direct quote otherwise you really need to just sort of read and understand and then write in your own words'. And I think what she was trying to do was rewriting sentences because I know her spoken English is very good, you know. She's, her English is fantastic, but I think her problem was that she was reading a well written sentence and then she was trying to change it to something else and it wasn't quite, yeah, wasn't quite coming across, yeah. (FS4)*

The difficulty in disconnecting language, academic discourse and understanding is made clear by this student. Her friend had 'fantastic' English yet was not able to re-write a sentence in a way that made sense. Her friend was attempting to do this because she had been 'told' about plagiarism and how to avoid it in a way that encouraged her to see it as a mechanical process, manipulating language as a technical exercise rather than as a way of demonstrating understanding of a topic. In order to enable students to move away from this mechanical view of language use, it is necessary to consider how language interacts with pedagogical choices within disciplinary sites.

Language and Pedagogy

Different knowledge structures shape social practices and forms of pedagogy (Bernstein, 2000; Maton & Muller, 2007). Therefore, different disciplines have different pedagogical approaches; a 'classroom' for the teaching of Plant Biology looks very different to a 'classroom' for the teaching of Political Communication, for example. Different materials and technology are used, different ways of communicating and building knowledge are employed and different assessments tools used. In this section, I continue to draw out themes from the data collected in the two disciplinary sites which exemplify some of these differences in pedagogical approach. How different the approaches are across the disciplines was highlighted through my observations. In preparing to observe the classes in the two disciplinary sites, I had decided to use the EAP observation forms used in the EAP teaching unit. My belief being that it would enable me to focus my observations on the language as used within a disciplinary context. Within two of the AHC observations, the use of this form was, to a degree, effective and I observed a number of teaching techniques that would be recognisable within a 'good' EAP classroom. However, in all the other observations, I simply found myself writing lists of lexical phrases, key terms, and a description of the interactions I observed.

Defining pedagogy

Essentially, pedagogy is the act of teaching. However, this act is loaded with epistemological and ontological assumptions and beliefs, many of which are being increasingly questioned to the extent of being described as 'deadly habits' (McWilliam, 2005: 2). Some of these are conscious and explicit, others are less so. Pedagogy is now a much-theorised concept within the field of education, and there is general agreement that any act of teaching involves an interplay of power relationships that have an impact on the learner in particular but also on the teacher. Pedagogical choices are made around whether to act as a 'sage on the stage', the 'guide on the side' or the 'meddler in the middle' (McWilliam, 2008: 263). There is no one approach to teaching and learning that is socially, politically or culturally neutral. As a result, much of the literature focuses on the transformative impact (both positive and negative) that pedagogical choices can have on the learners we work with (see, for example, hooks, 1994; Freire, 1996), with powerful arguments being made around how pedagogy can work to strengthen powerful hegemonies but also how, with intent, is can be used to undermine them. It is via pedagogic acts that a learner gains (or fails to gain) access to knowledge. In language teaching these acts are usually referred to as methods or approaches. There are a wide range of these methods, often strongly contested within the field. It is in the choice of method or approach to learning and the application of it within a learning environment that a particular pedagogy becomes visible. These actions can be self-directed; can involve interaction with digital content or can, more traditionally, be planned by another (the teacher or *pedagogue*) who is seen as having some level of expertise in translating, reproducing and recontextualising knowledge for learning.

Pedagogical acts are closely bound to choices made around the curriculum, and to an understanding of what knowledge is valid, how it is chosen, sequenced, paced and evaluated (Bernstein, 2000). In suggesting this focus on valid or legitimised knowledge within the curriculum, Bernstein's 'pedagogical device' was based on three fields of discourse – of production (new knowledge); of recontextualisation (knowledge placed in the curriculum) and reproduction (the act of teaching). It is this understanding of knowledge as central to both the curriculum and to pedagogy that seems to be most prevalent in Higher Education practices. This can be seen both in the recent push for research-led or research-based teaching (see, for example, Fung, 2017a, 2017b), which links the production of knowledge to the recontextualisation and reproduction of it, but also in the more 'traditional academic pedagogy, or rather non-pedagogy of osmosis' where 'the assumption of osmosis is predicated on sameness' (Turner, 2011: 21); teachers teach in the way they themselves were taught (and successfully learned), the assumption being that they are teaching students who are like them. In this way, it could (controversially) be argued that 'Bernstein's argument that the curriculum represents valid knowledge, suggests a highly predetermined curriculum that ensures that hegemonic ideas are replicated' (Bovill & Woolmer, 2018). The knowledge chosen to be recontextualised and reproduced within the curriculum may be powerful, but within that power lie specific socio-cultural assumptions.

Again, the increase and diversification in student population is having an impact on this traditional pedagogy and is highlighting the cultural and knowledge assumptions that currently underlie HE pedagogies.

Practice as a reaction to perceived international EAL student deficit

Across the two disciplinary sites, much of the pedagogy described and observed was based on individuals' own experiences of learning within the discipline, on reflection on individual practice, trial and error through practice and a sense that 'this is how I would like to be taught'. This understanding of pedagogy, developed through apprenticeship within the discipline seems, to date, to have been largely effective, or at least rarely brought into question. However, 'Pedagogy includes discourses about pupils – their abilities, motivations, in short, what kind of people they are – and these discourses influence teacher practices in the classroom' (Lefstan & Snell, 2014: 150), so with a changing cohort of students much of this experiential learning of how to teach seemed to be being tested.

The usual pedagogical practices in the two disciplinary Case Study sites were not as effective when students were seen as having lower levels of language proficiency and/or differing educational expectations. This was noted by students as well as teachers:

- *the teacher is saying, so you could start your discussion and we just talk in Chinese because we are all Chinese people. (6X)*

Thus, basic classroom management techniques, which are not the traditional concern of TPG teaching, needed to be employed:

- *I think the ... teacher is maybe more brilliant. Every group has one foreigner; not Chinese I mean [laughter]. (5A)*
- *Each time we had the topic she'd divide us four or five and we needed to talk and she will check. She will ask us. (6Y)*

Disciplinary teachers were also concerned with ways of ensuring active participation of all students in a class:

- *I think then that there needs to be, and I think more talking, you know, lots of talking in class, bringing people out rather than waiting for them to come in. I think not allowing them to get into that habit, where they never say anything. (M7)*

Others had considered different approaches to encouraging engagement and participation to ensure all students were included

- *We're providing a lot of online materials that are designed to walk them through the process a lot more. (M4)*
- *I open a Twitter account for the students to participate. So, the ones who feel shy can tweet and I can see and I read the Twitter in class. (M3)*
- *I also use the VLE a lot more as well again to try and get the Masters students confidence up to try and get their confidence a bit responding to ideas and responding to texts on the computer if they are struggling with confidence in the actual sessions. (M2)*

Although this focus on students interacting with each other and learning through discussion in seminars was not such a focus of the pedagogy in the STEM site, the need for sharing of observation and understanding remained clear. Thus, in one practical session I observed the teacher managing her class by regrouping students in order to encourage greater discussion. For the majority of the students, this classroom management technique was enough to engender greater discussion. For Mai, with less confidence, less expertise in English and less understanding, this was not enough to draw her in to more participatory learning opportunities.

There was very little criticism from student–participants of the *teaching* practices in either of the School sites, and students uniformly praised the support provided by their tutors. Comments made were around a lack of clarity over specific academic practices. These included vague assessment practices (as described in Chapter 5); a lack of understanding of the purpose of essays (something also expressed as an issue by staff, mainly in regard to the Chinese students); and wanting more seminar time. This supports the arguments made by those working in Academic Literacies that whilst student writers know they are expected to write within 'a particular configuration of conventions, they [are] constantly struggling to find out what these conventions [are]' (Lillis & Turner, 2001: 58).

In interview, teachers found it difficult to articulate their teaching expertise beyond providing examples of their own good practice in student education. This was almost wholly based on experience developed over time and through experience rather than related to any pedagogical theory or training. Many of the changes described were reactive, based on student feedback or requests. Some of these changes to practice related to the perception that students were struggling with language and needed more time to work towards an understanding of the content. In this way, the burden of understanding was again placed on the students, with expectations that they would spend more time in preparation for and review of the taught sessions:

- *I now always make sure that I've got my PowerPoints onto the VLE at least 48 hours before I do a lecture. I always used to put them up but sometimes immediately after the lecture. Or sometimes maybe just before. Now because of demand from particular students, then I need to get those up early because they then want to go through and make sure that they understand what I'm saying before I even say it. (M6)*
- *there's a lot more lecture recording goes on in the post-graduate programmes than in the undergraduate programmes. (M4)*

There was also a sense that you need to have particular attributes as a teacher that would allow all students to feel comfortable in the learning environment you create:

- *I've found them to be enormously competitive; more so than home students. When I do similar exercises where there's a competitive element, I can't shut them up. I would say organise every seminar you've got for the competitive dimension or in the early stages anyway. (M9)*
- *I've tried to have a chat with them and see where they were up to. Again, it can be a problem in the larger programmes because I don't do all the seminar teaching. (M8)*
- *you need some skills in getting them to work as a group and learning through discussion the hard way. You need somebody who's fairly lively really. (M8)*
- *appear to be an approachable enough person for them to want to come and talk to you about it. I've had a number of students, particularly Chinese students who come in and said look, I'm having real trouble with the language here. (M7)*

- *I speak slower. I try and articulate better, which is no bad thing really because the other students, that's fine. I do think the way of teaching has changed because I spend more time explaining academic concepts rather than moving on from there and doing the exciting thing of debating ideas, which I'd like to do a lot more of. (M6)*
- *If it's a seminar it's small and the teacher should be able to point directly to people and make them participate. So if you wait for them to stand up, they're not necessarily going to do it. You have to point and say, 'What do you think?' and listen and be patient. (M3)*
- *I always try and foster an environment where they're okay to talk to me and they're comfortable with coming to talk to me which they do which is great so if there's any real problems we can catch it early because my seminars are very informal, they're very friendly and I put myself out there as a really friendly lecturer so I hope they come and see me and they do, I've seen most of the students. (M2)*

Regardless of these changes to individual practice, it is clear that the overall pedagogical approach within the AHC site in particular is generally based on the assumption that learning takes place through dialogue, which is itself culturally bound up in assumptions of inherent, stable and context-independent abilities (Leftstan & Snell, 2014) and requires high levels of expertise in the manipulation and interpretation of language to build a coherent argument.

The student response to the current exclusion of large numbers of students that this approach creates was that more time is needed:

- *Because the seminar is just one hour I think, so I think it's too short ... Because the Teacher do have their ... they have their, something to say and so we think there is no time for us to express ourselves, yeah. (6Y)*

The teacher response is that more explanation is needed:

- *the way we teach and we deliver and we assess. In the UK, we think that it's very common to ask people for essays. That doesn't happen in every country. You have to explain to the students what an essay is and how you do it and what you explain in it. (M3)*

All of this takes up time, which, as has already been discussed, is lacking for all involved. I have included here some examples of good, inclusive pedagogical practice, but more needs to be done, and practices need to be shared through disciplinary collaborations. Without this, teachers continue to work in isolation, making reactive decisions based on immediate problems, rather than focusing on developing principled, evidence-based, inclusive practices that work to engage all students:

- *I'd be interested to hear from you if this is echoed with other staff. I haven't had the time to talk to colleagues. (M9)*

Chapter Summary and Practical Lessons Learned: Focusing on Language in Content Teaching

In this chapter, I have attempted to outline the participants' understanding of where language and content knowledge both intersect and can be disconnected. It is clear that, if language is considered at all, there remains for most a belief that it is a separate entity to knowledge and can be dealt with by an 'other' and separately. However, when asked to describe where the separation is, few were able to do so.

Language seems to have most impact and be most visible when there is a lack of clarity – from students in their writing and from teachers in assessment rubric, lectures and explanations. It is also problematic when it takes on specific disciplinary meaning that marks the user as a member of a specific community. Here the focus is on making discourse choices. Yet, despite being integral and inseparable, it is undervalued and seen as 'less'. Those students who need more support and development in language *proficiency* or in developing their academic *literacy* are inevitably placed at a disadvantage when working towards an assessment; when given help with this they are also then open to questions around the extent to which their work is their own – when does proofreading shift into collusion, for example? For this lack of value and lack of clarity around language practices to shift, for the language learning to be valued along with the content learning, it is necessary to take an integrated approach where language becomes visible.

There are some easy wins in terms of highlighting language and developing it through content. Here I provide a few examples of how teachers can begin to make language more visible in their content curriculum.

- Provide glossaries of key terms on the VLE. Ask students to add to and build definitions as their understanding of key concepts develops throughout the programme.
- Provide and discuss models of writing. Focus on what makes one example a better response than another. Do not only provide examples of excellence.
- Provide opportunities for different modes of communication. Use, for example, a class Twitter account to record responses; build collaborative lecture notes via a shared electronic document.

- Work in partnership with EAP teachers so that language and content knowledge are developed in parallel.
- Discuss language as an academic choice (see Caplan (2019) for resources on grammar as choice as one example).
- Draw on language as choice to consider how it is used to show a particular perspective in texts (see Caroll, 2015). Reflexively question the cultural bias or implications within language choice.

Again, these are a few simple suggestions of which there are many more that could be added. Other than the final point, they are relatively easy to implement as an individual. However, they will achieve very little if not part of a wider shift in understanding and a change in institutional culture around how language, language learning and academic literacy is viewed across the entire curriculum. I will address this more fully in Chapter 8.

5 Language and Academic Norms

Having considered how language impacts and intersects with knowledge communication and dissemination within the planned, taught curriculum, I now focus on the experience of international, EAL students across the entire curriculum as they interact with the often occluded institutional and academic norms and tacit expectations. Here I take a more holistic view, looking not only at language use within a disciplinary discourse, but also at how language connects with an individual's sense of (academic) self and can create or add to a sense of (dis)empowerment. Previous research, cited in Thomas and May, has found that not only curriculum design but also the 'hidden curriculum' frequently excludes certain students and privileges others (2010: 10). In internationalised contexts, where one language is afforded a higher status than others and is viewed as the 'norm' against which all communication is measured, language use and proficiency becomes intertwined with cultural and social capital. The ability to use that language to communicate complex ideas becomes an added 'threshold' (Meyer & Land, 2003, 2005) for students to cross.

Language and Cultural Capital

Cultural capital is viewed here as being held through the extent to which you are able to understand and participate in academic practices; the extent to which you are a 'holder of the legitimate competence, authorised to speak with authority' (Bourdieu, 1991: 69). This authority emerges from what Bourdieu describes as 'habitus' – our 'habitual patterns of understanding and inhabiting the world' (Zembylas, 2007: 446) – created through interaction with the world and social circumstances to become what Bourdieu also terms 'practical sense'. This concept of habitus can be used to explain how structure and perception interact to impact on our actions but also to explain how we work to survive in a complex world (see Bourdieu, 1991; Bourdieu & Wacquant, 1992). By developing a 'practical sense' or habitus, we learn to live within social structures without the need to constantly question or learn how to do so; the further developed our habitus is, the greater our legitimate competence within any particular field. It is this competence that gives us different forms of capital. Cultural capital, then, exists in three forms: embodied, objectified and institutionalised. In relation to the UK HE context, this can be broadly explained as relating to: the dispositions of mind and body – what are culturally acceptable ways of thinking and being (epistemology); what are culturally acceptable forms and objects of knowledge (ontology) and what are culturally acceptable qualifications (previous study; language proficiency). I would argue that discourse competence – how to write and speak in a culturally and contextually specific field well – cuts across all three forms of cultural capital. Others have also added to Bourdieu's conceptualisation of cultural capital, arguing that is should also recognise 'its affective aspects such as levels of confidence and empowerment' (Zembylas, 2007: 449). Many EAP centres have adopted the trope, based on Bourdieu and Passeron (1994), that 'academic language is nobody's mother tongue' in an attempt to suggest that it is not only EAL or international students who struggle to access the language of the academy and their discipline and to highlight the view that language development provides an opportunity for all to increase their cultural capital, for them to become empowered and more confident students. This (mis) usage has been recently critiqued by Ding (2019a) as dismissing the particular struggles involved in studying in a language that does not feel like your own; however, what is clear is that the ability to skilfully manipulate language to express understanding gives the user power that those who cannot lack.

For students to feel 'authorised to speak with authority', they needed to feel able to both interact with a text and with other speakers in unplanned communication both inside and outside planned learning episodes. The ability to do this is key to moving language learning beyond a language focused classroom, where language and skills are used in a more sanitised environment, to the more messy and unpredictable interactions of the academic world where there is little time for planning before the need to speak. The lack of cultural capital, experienced by international students as a lack of confidence or means to participate, was a strong theme across all three sites and was expressed clearly by students and teachers alike.

Student participants from the AHC site were able to identify this change in educational culture and concepts as an aspect of their academic journey very early on in the project (Bond, 2019). Thus, Focus Group 8, describing their depiction of their academic journey as enacted through the computer game Super Mario Bros. explained:

- *The flowers and turtle are the monsters. They kill or diminish Mario – the barriers. These are the readings and theorists. These are the difficulties we have to jump over in the game. It's easier to do the game as a national students, with the same cultural background. In China, we have a big class, the teacher says something and students listen. Here you need to talk and talk and talk. (8L)*

This chimes with the point made by Mai that UK students had different knowledge to her, and that this knowledge was better known and valued by her teachers.

UK students were seen as having access to, and greater awareness of, the specific discourses which have 'developed within disciplines to represent (and simultaneously privilege) particular understandings and ways of seeing and thinking' (Meyer & Land, 2003: 8). This view was supported by this student from the United Kingdom:

- *if I go back to when I was like 18, if I looked at a paper I just wouldn't understand it at all and you'd get used to it. And I think partly this is where all the basics come in you just know all the terminology and everything. (FS5)*

His experience contrasts with his international peer who accepted her tutor's point that she would now have to learn 'the basics'. In this way she had more obstacles to overcome in order to reach the same end:

- *I talked with my personal tutor and he said he understands, when I told him about my difficulty ... he said yeah, I understand because so many people from different backgrounds and it's fine, we're just getting you to do this so that you will learn ... when I see my classmates who were from England they are much, much better than me. (FS1)*

Students also made clear links to gaps in their language having an impact on their self-perceived cultural capital:

- *It's embarrassing sometime ... because for example, you are on the same seminar and you just come out with a very good idea but it's in Chinese and you put your hands up and then it's like ... I ask my Chinese friends. (C3)*

Thus, language is part of the 'privileged knowledge' that creates a barrier to effective (intercultural) communication (Ippolitio, 2007). The inability to express thoughts in an elegant, articulate and clear way 'diminished' students, exasperated them and, for some, led to a decrease in confidence. Students linked all of these problems together as an inseparable mass, touching on issues of national (self-)stereotyping and 'othering', language difficulties, the higher cultural capital that comes from the 'audibility' (Miller, 2003 in Ortactepe, 2013) of being a 'native speaker':

- *we have more than ten people in the group but only three locals. The others are Chinese. And every time the lecturer asks questions the three will discuss together and the other Chinese just listen. Language is really difficult. Sometimes we finish the reading but we don't quite understand it all and the local speakers someone talks. The other two sometimes speak so fast, faster than the lecturer, and it's really hard to chase her. And I think also it's a brave problem maybe because they give us too much pressure we just don't know how to say that. Or maybe we have the answer in our own mind but we're afraid maybe it's not quite good. (8M)*

Others, however, while agreeing there is a gap between L1 English and EAL speakers in understanding, which they believed resulted initially from linguistic proficiency, viewed it not as a problem but as an opportunity for learning and discussion and for increasing their own cultural capital:

- *In my seminar, I have colleagues who are native speakers but I think they speak better and they read faster so I think their understanding will be better than us. But I don't see it as a problem because it's very helpful when we discuss to each other. (5C)*

All the home students who participated in the project, none of whom were at the time engaged in language learning, commented on the respect that they felt was due to students who were taking a TPG programme in an additional language. Prior to participation in this project, most 'home' students from the STEM site had never considered the differences or difficulties experienced by international students in comparison to their own experiences. They were able to name students on their programme who were international but had not used this as a particular means to separate them out, nor viewed them as needing extra support or being in deficit in any way. In contrast, the home students in the AHC site, being in the minority, had clear but very hedged positions on international students, again more specifically Chinese students.

- *you can tell the difference between speaking with your native and speaking with an international student, you can tell the difference but it's just about patience I think for me when it comes to working with my group. (C2)*

More concerning was the stark appraisal from teachers around the collaboration between Chinese and non-Chinese speaking students on the programme. Whereas C2 suggests the need for patience, and her main seminar tutor also suggested that relationships between students were good:

- *I've never noticed any kind of antagonisms or any tensions at all, they forge quite close friendships. (M2)* other teachers suggested a much bleaker picture:

- *a lot of the students in the class who are not East Asian don't have the patience either, and so there's frustration. That can lead to all sorts of rather nasty things which we don't want to happen because these students are so keen to learn and so bright that they don't deserve that kind of treatment. (M6)*
- *that impacts on students from other parts of the world who get very frustrated in seminars because they've read the material, can discuss it, and partly because of culture, East Asian students don't particularly want to contribute, and partly because they haven't understood, or sometimes haven't read because they've given up, the material before the seminars. So that then impacts on all the other students. (M6)*
- *The Chinese students ... feel segregated. They also feel it doesn't help them academically because they feel they're getting one kind of tuition that's a lesser standard perhaps than the others. The others have got rather superior and they are now complaining that they don't want to be put back with the rest of the Chinese students because it slows them down. (M6)*

This situation, then, is the current lived reality of staff and students who learn and teach within a context where there are a majority of students who come from outside the 'home' language and cultural learning environment. This is particularly clear when a large number of these students also come from the same first language background and culture and thus form a majority–minority. It is a situation which is difficult to discuss as it can very quickly move towards essentialist language or worse, largely because the issues are so complex, involving intersections of identity as well as issues of power. Theories of culture become employed by different groups to construct ideological imaginations both of themselves and others leading to the demonisation of a particular foreign other (Holliday, 2011: 1). A great deal of care needs to be taken around discussion of the complexities involved. However, without providing the space for open dialogue around the current difficulties faced by teachers and students alike, HE institutions will continue to do all a disservice. Again, I suggest that this is an area where EAP has a role to play in conjunction with the growing institutions for teaching excellence and scholarship, acting as a critical friend to colleagues in other disciplines and encouraging a focus on practical pedagogical responses to a changing cohort of students that draws on EAP and intercultural research.

There is a clear danger in focusing on 'culture' in the context of internationalised education. Holliday (1999, 2010, 2011; Amadasi & Holliday, 2017) has written much about the need to move away from thinking in terms of 'big' cultures, where individuals are categorised by nation state, ethnicity or religion. Intersectional studies have also highlighted the phenomenon of individuals (whatever their nationality or culture) feeling the need to 'represent' the group(s) they identify or are identified as being a part of in some way, and how this places a burden of responsibility on them. Gao (2011) describes how the participants in her study of Chinese study abroad students felt that they became 'more' Chinese and needed to provide a positive representation of Chinese people to other students and teachers in their language school. The tendency to other and self-other in terms of culture and linguistic ability was a strong feature emerging from my data and is highlighted by the surprise expressed by this teacher:

- *this I don't understand, in any one cohort if I were to look at the top five, there'd be at least one Chinese student in there. (M6)*

It would seem that in this site, Chinese students have become synonymous with low levels of academic achievement and lack of social and cultural interaction to the point that it becomes surprising when one student bucks this trend.

Within all of this, it remains unclear whether the core problem lies in linguistic competence or educational differences and where (and what) culture intersects. I would argue that in this context the two are so closely intertwined that they cannot be separated. As Montgomery claims:

> Learning culture is similar to learning a language, and, just as language defines our ability to understand and explain what we observe, so culture is a means of framing both propositional … and procedural … knowledge of that which surrounds us. (Montgomery, 2010: 8)

There are instances, for example in English as a Medium of Instruction (EMI) context or where a Canadian is studying in the UK, where language and culture do not tightly intertwine. However, a large number of international students find that not only do they not have the linguistic resources to fully understand, nor do they have knowledge of the propositional and procedural framework they are expected to work within. The combination of the two works to remove all self-efficacy within the academic context.

However, just as teachers maintained that they were able to separate language proficiency from disciplinary understanding, there was also suggestion of a separation between language and culture:

- *it's not always a language thing, that's also to do with wider cultural questions. (M7)*
- *but it feels like it's not the language, per se, that's the sticking point but cultural practices of learning. That's the narrative I'm starting to tell myself. (M9)*

The culture referred to by these teachers was specifically around learning and educational practices, which many concluded was the main barrier to students being able to participate in the learning process in the United Kingdom:

- *When I speak to them individually, and in the seminars as well, their English is good. I rarely get the impression that their language is so poor that they're having problems … The difficulty of engaging with the material often seems more a matter of cultural habits, I suppose, around the learning process and what it means to engage with the learning process. (M9)*

This suggests a view of language competence as being measured against the ability to take part in an individual conversation. This idea of the level of language required is echoed by this teacher, who felt that the ability to communicate via email would be enough to meet requirements:

- *to be able to contact me either by email or meet with me and speak with me if they've got a question and for us to have a conversation and maybe if I'm able to answer their question for them to get a favourable response. I think that would be the basic requirement and I think our students far surpass that. (S1)*

In fact, despite the suggestion that educational culture might be a more important factor in student learning, language was never far away from cultural considerations. Here the move from thinking about all international students, or all EAL students, towards a specific focus on Chinese students again becomes most obvious.

- *There is a difference but it's not necessarily a linguistic one, it is a cultural one. I mean Chinese students in particular … East Asian students, and not just Chinese but probably predominantly Chinese students … They don't get the idea of critical analysis and forming an academic argument in the same way as we would. So, I think students from those areas have a double problem. The first is that I don't think their English is as good … The other reason is because of the culture of the kind of theoretical study that they do. So, they have this double problem when reading articles in academic English. (M6)*
- *It's nearly always Chinese students who have that problem … when it comes to the sort of semantic core of the language, it's the Chinese students. (M7)*

Where Chinese, and other international students, were seen to have *greater* cultural capital was in their work ethic and the extrinsic expectations placed on them by family:

- *Nearly all of my students are very, very hard working, very dedicated and very good students and so I can rely on them, in their own time, to go over and over and over again, to proofread, redraft and to do that work. (M8)*

- *the bulk of our students really are very, very good and so they're coming from backgrounds where they expect to perform highly. (M8)*
- *In fact, they're better than the UK students when it comes to enthusiasm, punctuality, desperation to learn. There's a limit to what we can do when it comes to going back to start at square one, but we do a bit. (M6)*

These conclusions continue to place expectations on these students as a whole group to work harder in order to achieve the same results as the more traditional TPG student. These expectations do not work to include, but feed into a culture of self-help and the need to work harder than others in order to achieve. Turner (2011) suggests that some of this perspective around work ethic comes from a Western interpretation of the meaning of the often used 'I will work harder' at the end of a tutorial, which she proposes is used by Confucian heritage students as pragmatic signalling of politeness rather than a true commitment to improved work quality. Assumptions of greater work ethic and a placing the burden of extra commitment on EAL students is an approach that will ensure the continued isolation of students who need to spend more time on independent study than others and could ultimately result in mental health difficulties.

Of particular concern here is the assumption that it is the international, the EAL, the Chinese students who are the cause of the problem. Kettle (2017) argues that the perception of international students as operating in deficit is no longer accepted as being worthy of consideration. Yet this seems to be the case more in the research literature than in the current practice, experience and perceptions of teachers – at least in the context of my scholarship work. Teachers and students were aware that expectations, practices, culture, language, epistemologies and ontologies appeared to be misaligned. Participants were able to reflect on this, consider reasons and discuss possible solutions, yet there remained a firm belief that current cultural practices and the paradigms the sites were operating under were legitimate , and that it was the work of the students to overcome their deficit and to integrate and become audible, as defined by Ortactepe (2013: 223) as 'the degree to which speakers sound like, and are legitimated by, users of the dominant discourse'. There remained a sense that students should be inducted and socialised into, rather than transform the current academic and disciplinary culture of each site. The power remained within the current norm:

- *I'm assuming they've come here because they know what they're doing and then we try to integrate them as much as we can. (M7)*

Transformation, seen in Critical Realist, morphogenic terms, does take time. Current curricula and higher education practices have been in place for long periods of time. International student numbers, especially from China, have increased exponentially over the last decade, causing a need for a sea change in approach. There was evidence of the beginning of a transformation, and of the cultural capital of being an international, EAL student beginning to be valued. Site 2 as a whole had clearly entered into discussions around how to engage and include their students, and were beginning to consider the content of their programmes as well as their approaches to teaching as a key element in overcoming some of the cultural barriers they perceived:

- *I think we've become a lot more aware of the student experience. I think we have de-Westernised some areas of the curriculum. (M8)*
- *The, you know, content is tailored to the students. It is designed to engage them with things that are both challenging and familiar so there is a lot of Chinese media. (M4)*

This member of staff in particular suggested that there needed to be a move towards a more fully globalised curriculum, and shifted the problem away from the students:

- *I think the problem is not then with the students. The problem is that we don't realise enough that this is a globalised world in which we have to adapt our teaching to the realities. (M3)*
- *I think that's the problem that we have, not the students, that we need to make ... this more global cosmopolitan approach. I think it's a problem of UK universities in general that we tend to deliver this teaching in terms of a packed product that is universally applicable in every person. It's not the case. We need to realise that it's very different in different countries and our teaching and our delivery should reflect those things. Both our curriculum and our syllabus should reflect that diversity. I think not only in teaching but also in enriching the learning of the students but it also makes us better teachers. (M3)*

The implications, therefore, of the levels of cultural capital, understanding of academic processes and norms, and the linguistic and culturally biased metacognitive ability to gain access to developing this capital is already having pedagogical impact in areas of the University where there are the greatest number of international students. This suggests a need for strategic curriculum change both in order to prevent a two-tier system from developing but also to harness and celebrate the cultural diversity that an internationalised cohort of students can bring, adding to the cultural and intercultural competence and capital of all students rather than positioning some as culturally powerful and others as culturally deficient.

Language and Social Capital

Have you ever worried about the other international students who are different from China, not from China, but may be isolated? (8P)

Social capital, in Bordieuan terms, exists as and through social relationships, networks and groups yet belongs to the individual. The networks and relations both act as resources that members are able to mobilise to their advantage but also work to create and reinforce community norms and social structures, encouraging conformity amongst its members. Social capital encourages a collective identity that 'enables individuals to feel themselves part of a community as **opposed to others that are considered to be outsiders**' (Zembylas, 2007: 457; emphasis added). Extending an understanding of social capital, particularly within education, others have suggested the connected concept of emotional capital (Zembylas, 2007; Cottingham, 2016), defined as the emotional resources, such as support, patience and commitment, that are built up over time and through social relationships. As with social capital more broadly, emotional capital also unites through the exclusion of those outside a specific community – invoking emotions of fear, distaste, rejection towards the 'others'. Social and emotional capital are therefore inextricably linked to structures of power and privilege. In a UK HE context, this social capital remains with the white, English-as-a-first-language, 'traditional' academic or student, a group which maintains its status as the embodiment of a broad community that 'others' aspire to become a part of; membership apparently denoting acceptance into 'the academy'. However, the reality of students in some sites of my study, particularly in terms of daily social activity and emotional support, suggests a slightly more complex situation, with the possibility of slow morphogenesis as different individuals interact and act within social power structures.

The separation between Chinese and other international/EAL speakers is once more highlighted when looking at language as social capital within the TPG experience. However, this was only the case in the site where Chinese students occupied a majority–minority position. In the STEM site, where there were only two Chinese students on the programmes I had contact with, the same patterns did not emerge. In fact, these students seemed isolated socially as much as academically, with 'home' students continuing to hold the most social as well as cultural capital.

When in the majority, with no requirement to use English, Chinese students expressed a sense of empowerment and choice, which separated them as a group from the rest of the student cohort. Duff (2012) has described this kind of choice as a form of resistance to cultural norms encountered in the learning environment, connecting agency and power with social context.

- *I prefer to study with foreigners but when I go back to my room I like to play with Chinese people because we have the same lifestyle so if I live with foreigners maybe they do something I don't like. (6Y)*

This had not been the student participants' initial intention or expectation, but very rapidly became an expressed wish as they realised the difficulties of communicating across cultures:

- *We want to socialise with people from different countries but it's hard. Yesterday we had a Welcome Party. Western students sit together. Chinese students sit together. People still play with people they are familiar with. (8L)*

As in Gao's study (2011), the Chinese students reported increased awareness of their being Chinese and of a need to positively represent their country through their actions. They chose to spend free time with those who would not be judging them as the embodiment of a nation and to whom they did not feel the need to explain their behaviours. International students in general also rely on each other rather than on UK students to learn how to live in the United Kingdom, and build a social network that is supportive and powerful (see Montgomery (2010) for further exploration of this)

- *The 2nd part of the [Super Mario] poster: native to foreigner, looks at social identity. It includes obstacles of different habits of eating or drinking – food difficulties we must get over. Knowing other Chinese helps us to find places to buy food, supermarkets. We are still playing the game. (8L)*

One of the main reasons given for this choice of social group is having little emotional or cognitive resource to spare from academic study. Students were aware of the lack of time available for TPG study, and the improbability of developing lasting friendships, so initially made social choices based on ease of communication. The lack of desire to socialise beyond their own first language community, whilst empowering in many aspects, also seemed to have an unsurprising negative impact on students' language development. Thus, a Chinese student towards the end of the project reports:

- *... actually, I do feel very nervous when I talk to local people because I think my vocabulary is not good enough, it's not sufficient so every* *time a local maybe, talks to me, I just, don't talk to me, I don't know what to say, but now maybe I can maybe make small conversation or a discussion with him or her. (6X)*

This is in sharp contrast to a non-Chinese international student who had had little contact with others from her own language background, speaking very early on in the project:

- *at the beginning, I was afraid that I would be misunderstood when I speak because sometimes the language does not help me to deliver what you want to say. I found that it's very interesting to have friends from different cultures and I enjoy them. I'm learning from them a lot. Yeah, we're becoming closer now, I believe. I think that is makes me also more confident that it's okay to make mistakes and it's okay to just practise and it will be better with time. (5B)*

Yet even those who did attempt to participate in wider University social opportunities found it difficult to manage their time adequately to combine social and academic activities:

- *I really think it has many opportunities to do everything. I found it's full of essays. All assignments. We just finished one essay and the other is come. So, at the beginning of the semester I go to some societies and enjoy them and I found I don't arrange my timetable correctly so in the middle I hurry, hurry, hurry about my essay. (5C)*

By creating strong and powerful monolingual social groups yet continuing to see themselves as lacking in academic cultural capital, relying on others to speak in seminars, the potential for exacerbating the issues touched upon by the content teachers outlined earlier is clear. Thus, the home participant in Site 2 described her own sense of social isolation as a result of being in the minority:

- *I wanted to drop out the first week because I thought it was going to be a bit too intense for me and there was no one on the course that looked like me or anyone that was British either so I was around more Chinese students so I don't know how that's going to work ... still a lot of the Chinese speak in their own language and clearly I don't understand what you guys are saying ... I think it's not having solid friendships, sometimes I feel like I'm by myself. The course itself, doing the work and everything is fine, but it's just being able to socialise with people and my study group is fantastic and we've already went out for dinner but obviously it's just different because only British student, I think there's one other one too ... (C2)*

Norton Peirce (1995: 13) suggests 'it is through language that a person negotiates a sense of self within and across *different sites at different points in time*' (italics added); as the year progressed, and understanding of academic requirements became clearer, many students experienced a change in attitude, wishing to develop social relationships with those outside their own L1 community. Generally, however, as Lin explained (see Chapter 2), they found that by this time it was too late.

Although similar questions were asked of student and teacher participants, only students made any comment about the social aspect of differences in culture and language proficiency. From a teacher's point of view, then, the social curriculum was seen largely as separate to the academic. Teachers did comment on the various opportunities that were available for students outside the taught curriculum and felt that information around these opportunities was clearly provided. Some focused on the levels of pastoral support that they provided when they noticed a student was feeling isolated:

- *It's very important to have that conversation early and for us to take the responsibility to do so because it's quite easy, when you're teaching, to say my two hours is over, I've done that session, you know, I'm quite tired after doing it and not to go away and just do that extra bit of thinking about what's going on with that student over there in the corner, why are they like that? At which point you, I would normally, and I think this may be unusual, maybe not, I write them an email and I just say why don't you come and see me. And when they come in, I say how's it going? And nine times out of ten, the ones that I've selected who are like that have got a problem and they haven't been proactive about telling anyone but when you ask them, it all starts coming out.* (M7)

However, this was not systematic and was in fact seen as 'unusual' and positioned as better or 'more than' normal expectations and therefore not available to all. Furthermore, the sense of students needing to fill the gaps, and of operating within a deficit model continues into their social practices, with teachers suggesting they were making unwise choices by socialising within their own language communities:

- *one of the things that I, you know, I know about the way these students live when they're here or some of them at least is that they, you know, they know what's good for them and then they still don't do it and that's hard and that's hard for staff. ... We cannot keep ramping up the acceleration forever and I don't think it's a question of resourcing not being there but I do think you also want to understand the motivations of students because our resourcing is very often not used all right. We are making stuff available to our students and they are not using it so it's not a question of capacity. ... There's motivation, there is what is their purpose as a student? Why are they here? Why did they choose to come here instead of staying home? (M4)*

It is ridiculous to suggest, as a Director of Graduate studies at Duke University, USA infamously did in January 2019 (The Guardian, 2019b), that all students should speak English at all times in the 'corridors'. Yet the global cultural capital that is currently gained from being able to speak English is one of the reasons that students make the decision to study in the UK. It is therefore necessary that institutions do at least provide them with real opportunities and resources that enable them to participate and use English outside the classroom if they wish to do so. As Lin suggested, some of her peers' motivations for coming to the United Kingdom were not strictly academic at all, but were more around exploring the world

- *I think that because we have different object, some people come here like just for a vacation ... And some people are here to learn something new.*

However, neither academic nor social English language is likely to be developed if students find themselves in the situation described here:

- *Most of the class mates are Chinese. Sometimes, I just feel like I study. Last week, in a lecture, I just saw all Chinese there and I forgot to speak English and I was just speaking Chinese. (5A)*

What became clear in my study was the difference in perceptions of where social interactions began and ended. For international students, socialising took place mainly within their student accommodation and was seen as distinctly separate from academic life and work, which was essentially carried out alone. For home students, the lines were far more blurred, therefore creating far more unofficial learning spaces and developing supportive relationships that enabled the development of collaborative and deeper understanding within a social context:

- *I always come into uni to work. And I always go to the computer room so there are other people around. Just so that you can ask them things and bounce ideas off them and things like that. Because I think that is massively useful to be able to talk to other people. Even just like, there can just be like some little thing you don't understand and they can just help you with that or you get ideas off them. (FS5)*
- *I try to stay here when I can because when I'm here I'm focused on work. There's too many other distractions at home ... it is a difficult course and, you know, everyone struggles to some extent so there is some sort of communication on helping each other as well, which is nice. (FS3)*

Even when international students were aware that their peers were working together in this context, because of their social circumstances they often chose not to participate. Accommodation and food options therefore impacted on social interaction and the learning that went with it:

- *They seem to just mix with everyone normally. I think when did – like maybe at weekends and things they stay more within international community because you see just like on Facebook or whatever then XXX will mainly be doing things with like international students at weekends. But I think that's partly just because they just get put in halls. (FS5)*
- *so I know two who live in halls really close and they just go home after every lecture so whereas it's half an hour for me to get home so I'll come in at nine and I won't go home til five so I'm with my friends all day so it's easier to make friends that way whereas they'll go home and not use the university computers so much. One of my friends is from Mexico and she's really lovely. She's often in the computer room with us, doesn't always like come for lunch or everything even though we would ask or whatever but just doesn't want to. And apart from that I don't really see them. I only know them 'cause they're my closest friends and I only ever see them when it's kind of timetabled hours I guess. We don't have many socials on the course. (FS2)*

A further reason for the separation of home and international students as they studied outside the classroom was the more obvious one of ease of communication and therefore the speed at which discussion could take place (see Mai in Chapter 2). This was also the case when students were expected to participate in group work, with one student reporting how she worked and communicated with other Chinese students first before 'translating' for their English group leader.

Therefore, social isolation, social empowerment and social choices all impacted on students' ability and desire to participate in their academic studies, and particularly on their continued language development and use. Being intertwined with emotion, the social capital that participants felt they had contributed to both their resistance to and acceptance of the need for a shift in perception of how language carries meaning and is not just a transparent academic subject. Social capital, therefore, impacted on the extent to which students were able to cross a number of language connected thresholds.

Language as a Threshold Concept

The notion of Threshold Concepts was developed by Meyer and Land (2003, 2005) as a means of describing and identifying the most 'troublesome knowledge' that learners encounter (Perkins, 1999). A

Threshold Concept describes knowledge that is 'alien, or counter-intuitive or even intellectually absurd' (Meyer & Land, 2003: 2) and takes time, effort and often discomfort for a learner to grasp. Meyer and Land describe the process of grasping these concepts as being in three basic stages: pre-liminal (not knowing); liminal ('a suspended state in which understanding approximates to a kind of mimicry or lack of authenticity' (2003: 10)) and being across the Threshold. These stages are not necessarily linear; learners may find themselves 'stuck', particularly in the liminal spaces where there are multiple layers of almost knowing and where a mimicked understanding may become fossilised. However, once the Threshold is crossed it is probably irreversible. Once a concept has been grasped, it is impossible to go backwards to a place of not understanding, and the knowledge becomes, therefore, tacit, ritualised and bound into an individual's identity. As such, any Threshold Concept is not only conceptually difficult, it is transformative and can, therefore, be personally disturbing, leading to strong resistance to the concept in some learners.

As a means of identifying complex knowledge structures, it is unsurprising that Threshold Concepts are also bound up with language use. As Meyer and Land state:

> It is hard to imagine any shift in perspective that is not simultaneously accompanied by (or occasioned through) an extension of the student's use of language. Through this elaboration of discourse, new thinking is brought into being, expressed, reflected upon and communicated. (2005: 374)

Language, then, develops alongside the developing understanding of disciplinary discourses. Yet language itself is also troublesome because 'Specific discourses have developed within disciplines to represent (and simultaneously privilege) particular understandings and ways of seeing and thinking' (Meyer & Land, 2003: 8). Threshold Concepts are, in this way, closely connected to the notions of Cultural Capital discussed earlier in this chapter, where the concepts and language chosen are bound in cultural and disciplinary norms, thereby reinforcing power relationships within the curriculum. Those who are fully socialised into the norms of the UK Higher Education system, who have successfully crossed disciplinary Thresholds and successfully communicated this understanding in previous assessments to achieve their own academic qualifications and through academic papers, have also crossed the Threshold that leads to ways of thinking and practising, all of which are 'signified' through language use. As Hall suggests:

> To speak a language is not only to express our innermost, original thoughts, it is also to activate the vast range of meanings which are already embedded in our language and cultural systems. (1996: 608–609)

In this case, the language and cultural systems are those of the discipline. Once this language becomes embedded, and embodied, once the Threshold has been crossed, it also becomes tacit. The language, discourses and literacies of a discipline are, therefore, difficult for disciplinary teachers to clarify as they have become a part of their identity and their 'habitus', but are also complex and troublesome for students, and even more so when the language is itself 'alien'.

Language as Tacit Knowledge

Understanding and knowledge of language and the role it plays in academic content communication is, therefore, opaque and troublesome. It is in the conceptualisation of language as a Threshold Concept that cultural and social capital again become key. Just as one clear distinction that can be made between a native speaker and a learner of any language is that the native speaker has no real memory of consciously learning the language; they have no conscious awareness of how they moved from not knowing to knowing; equally, those who have become expert users of their disciplinary language (regardless of their home language) and discourse also seem to have little conscious awareness of how they learned the language of their disciplinary tribe; the language is a part of their habitus that is deeply unconscious. While they may not have been born speaking academic English, their social, cultural and educational capital frequently ensures that they have been able to gain access to it with relative ease (Bourdieu & Passeron, 1994). Here it is not the language you speak at home that allows or denies access but the social and cultural background or habitus you occupy. In this way, then, language, and specifically disciplinary language, provides or denies cultural capital. By viewing academic/disciplinary language also as a powerful Threshold Concept it also becomes visible as a knowledge-based concept that requires carefully scaffolded teaching in the same way that any other more commonly visible knowledge-based concepts also require scaffolding

Among my participants, there was both a struggle to understand and a difficulty to explain where any understanding came from once it had been gained. Teachers need to attempt to travel backwards over the threshold and learn to see the language and discourse used in their discipline through new eyes; through the eyes of a student who may not speak the same first language; who may not have the same educational background or cultural understandings and norm references. Consideration needs to be given to jargon, conceptual language and cultural references throughout the curriculum. Teachers also need to develop a full awareness of when language and when disciplinary knowledge is the issue in the work a student produces. Quite often, the two seem inseparable:

- *I can imagine people getting, some people at least, getting quite confused about where the line is between bad grammar and nonsense for example. (M7)*

It is, however, vital that all teachers who work with all students begin to view language as more than just a means for basic communication. There is a need to shift from the linguistic expectations around basic communicative ability, such as being able to write an email, towards acknowledgment that the nuanced discourses of disciplinary experts can create barriers to learning for (not only international) students but also that these discourses need to be explicitly taught – either within and by the disciplines themselves or in collaboration with EAP practitioners. Acknowledgement that their own disciplinary discourses and ways of thinking have become part of their habitus, is the first step towards understanding these barriers:

- *I'm constantly struck by the difficulty of explaining critical analysis to students ... because you just kind of know when you're doing it. ... So, to deconstruct it and think about it differently is quite hard. (M10)*

This shift moves all those involved in Higher Education towards an understanding that language and academic literacies therefore need to be explicitly taught as part of the curriculum. In order to do this, it is necessary to be able to breakdown and analyse, and then systematically re-build, disciplinary language in the way that Threshold Concepts research suggests should be done for other forms of knowledge. This is exactly what EAP practitioners should be doing within the disciplinary context (see Chapter 7 for discussion of the knowledge base needed to do this).

For students, the journey over the language threshold is even more complex, covering learning on many different planes. Over the timespan of this project, what emerged was a picture of students beginning to view the purpose of language and language learning differently. On arrival, and during their pre-sessional EAP learning, most students viewed language in two ways. The first view was of language as an instrumental hurdle for them to jump, that would allow them to begin the more interesting work of studying their own discipline. Once the pre-sessional programme was over, and they began their TPG programme, there was a general feeling that language learning would end; that the IELTS score and extra language learning on the pre-sessional was preparation enough for their study.

The second view was of the learning of English language as an academic endeavour, done in isolation of other forms of study;

vocabulary and grammar should be taught and learned but there was little sense that this learning had practical application beyond the language classroom. This is very much connected to the 'underpinning psychological and ideological elements of the IELTS test [which] are grammatical accuracy, spontaneity and flexibility' (Short, 2012 in Pilcher & Richards, 2017: 9). Students seemed unable to match the process of language learning to the practice of language in academic use, particularly to the unpredictable and messy practice of real communication. Bruce (2008) argues that the context in which we have experienced objects or properties is deeply connected to the meaning we give to words; classroom language learning places different meaning on objects to those learned and experienced in a laboratory(see, for example, the words discussed as jargon in Chapter 4)

Thus, there was a necessary shift from the end of the pre-sessional, where much of the student feedback focused on learning grammar already studied, for example:

- *It would be better if we could learn more about English grammar, which give nearly all Chinese students a headache. (anon)*

towards a more complex sense of struggling to express difficult thoughts through a foreign language

- *in many, many places, the grammar, the word, my thoughts, I think is not very ... I don't know how to say in English. (6Y)*
- *maybe sometimes we will talk about some deep questions and use academic words and maybe sometimes we can't express the real meaning by English. (6Z)*

Students expressed this shift in perception of the real purpose of language but found it problematic to enact in their own reality. They had begun to disregard their 'existing understanding of the signifier in favour of a new one for the new context' (Land *et al.*, 2014: 204), but did so at different points in the year and to different extents. For some, the transition was deeply troublesome, and the affective noise created resulted in a sense of being stuck. Many continued with a desire to learn more grammar, while acknowledging that they had already studied grammar for many years in their home country; there was a continued reliance on translation but an emerging understanding that translation did not provide enough depth of understanding. One of the content teachers suggested that this was a lack of understanding of the complex nature of their discipline:

- *For them, that may just not register as these are areas of investigation and not simple labels that attach to thing. (M8)*

However, this can also be seen as a form of resistance to the identity change and transformation that comes with understanding language learning and use differently, compounded with the identity shifts involved in communicating in a different language more generally. Kramsch (2009) writes of language learning as being a total body experience; learning EAP occupies both the physical and cognitive realm to extreme levels. Academic disciplines use language, technical terminology and methods of arguing and expressing thought, knowledge and understanding that are fully bound up in the culture of that discipline but also the culture of the country in which the teaching is taking place. The words used represent 'the inert crust' (Voloshinov *et al.*, 1973: 48); a successful student needs to move beyond this crust to understand language on an almost physical level of 'just knowing', of feeling something intuitively and being able to 'live' in this language (Tawada, 1996 in Kramsch, 2009) so that it becomes a part of their habitus and they are able to put all of their efforts into furthering their understanding of their subject. For students who do not have immediate access into the language of UK academia this is 'troublesome' to the extreme, as the two students below attempt to express:

- *you have a mass of things in your mind, I know this, I know that and you just don't know how to express it out, because sometime you need an English word but you only have a Chinese word and sometimes they reverse'. (Lin)*
- *I think I have two or three weeks during the semester I really – I don't know what happened. Just hard to understand people's talking and also hard for me to speak my own ideas. And I talked to my friends and it just suddenly go well ... I remember those weeks were really hard in the lectures to understand and talk to my friends. Sometimes I just didn't know how to express. ... And suspicion upon myself at that time is really difficult. But I don't know, maybe the language is not so stable. Maybe this situation will happen again. (5D)*

Awareness that the 'language is not so stable' is, I would argue, part of crossing the threshold into the liminal space where real learning can take place. For many students, though, the complexity involved in the intersecting pulls on their identity, and the choices they needed to make around where to focus their cognitive efforts can mean that either through resistance, lack of time or lack of resources, they remain in linguistic liminality throughout their time on the TPG programme.

Chapter Summary and Practical Lessons Learned: A Language Connected Curriculum

In this chapter I have suggested that language knowledge and use unsurprisingly permeates and impacts on all aspects of a student's experience and can lead to a feeling of powerlessness.

As language is inextricably linked to culture, both describing, revealing and building cultural norms, those who do not have, in this case, English as part of their habitus find themselves lacking in cultural capital and feeling that access to the cultural norm is blocked.

Within the social realm, however, there is a more complex picture. EAL students who form a majority–minority are consecutively socially powerful yet lacking in capital. Their collective support meets emotional needs and provides care. The homogenous 'otherness' excludes when there is a desire to be included, compounding a lack of cultural capital.

Emotional reactions to inclusion and exclusion in turn have an impact on students' ability to learn, on their willingness for or their resistance to continued language development. In this way, language is integral not only to students' ability to access their curriculum in terms of being able to read, listen, write and speak about their discipline, but also in terms of limiting their power. In feeling socially, culturally and at times emotionally excluded, they find they are unable to make effective use of even the language they do have. Reactions to the identity shifts and transformations created by this status sometimes enable a student to move across linguistic thresholds but often ensure they remain stuck in liminal space of mimicking but not fully embodying a new discourse.

Helping students to cross this threshold is not something that experts in other content areas can be expected to do alone; it takes clear expertise in an area to be able to breakdown genre, discourse and language from within a disciplinary context. Again, then, it seems clear that there is a need for EAP practitioners to work alongside those delivering content knowledge in order to enable students to cross the linguistic threshold of language use for disciplinary academic purposes.

Here again, in summary I provide a few practical steps that can be taken in order to address some of the issues covered in this chapter.

- Connect academic and social support together more explicitly. Highlight the importance of both to learning.
- Build informal learning, for example outside class work in computer clusters or other shared common spaces, into the explicit curriculum. Do not leave it hidden. Make it an expectation rather than an assumption (see Bleasdale & Humphreys (2017) for further recommendations on this).
- Focus on the resources students have rather than those they don't have. Avoid creating an environment where students feel that it is shameful for them to use a language other than English to communicate and express their thoughts. Find ways to embrace this extra resource and bring it into the formal learning environment. Can you establish collective agreement with your students around when different languages could be useful and when it might also become exclusive? (see Canagarajah (2011) for further suggestions).

- Accept and teach effective use of translation software. Ensure institutional agreement on this so that all students receive the same message (see Mundt & Groves (2015, 2016) for further discussion on this).

I accept that there are problems inherent in the official acceptance of the use of other languages and in the use of translation software, and that questions might be asked over the extent to which students should be supported with their work outside the written curriculum. However, these are all practices that take place already, but do so in the shadows without ever being discussed openly. By not debating their use, effectiveness and drawbacks, by ignoring 'the problem' and therefore implying they should not take place, students feel dis-empowered, guilty or ashamed yet continue to engage in these practices. By not being explicit in expectations and our understanding of what is acceptable practice, the line between real questions of Academic Integrity and the normal academic practices of peer review and feedback becomes blurred. It is in engaging others within the academy in these discussions and suggesting alternatives to current practices, based on scholarly understanding and teaching expertise that EAP practitioners can have impact, move away from the margins of academia and shorten 'the gap between what is and what ought to be' (Ding, 2016: 13).

6 The Place of English for Academic Purposes

As yet, I have barely touched on the position or role of EAP within the curriculum nor on their interactions with EAL students, academic teachers and disciplinary content knowledge. I now do just that. In this chapter, I consider the same four themes of identity, agency, time and trust as I did in Chapter 4. However, now I position disciplinary language as central to the curriculum and as, essentially, woven throughout all knowledge building and communication. Language is also positioned as a barrier to this knowledge building and communication, not only, but very particularly, for those students who are studying through English as an additional language. It is also a barrier to the wider and hidden curriculum, meaning that international students frequently find themselves missing out not only on the traditional learning experience but also on the social and cultural interactions that make up the wider learning experience of being at a University. They also often struggle to access support structures, which can lead to a heightening of issues around physical and mental health and wellbeing.

It is the responsibility of any university to ensure that all students are included and have equal access to all aspects of the experience it provides. It therefore follows that it is the responsibility of that institution to ensure that students have access to support and development opportunities that focus on the academic language and literacy skills that they need in order to play a full part in university study and life. Currently this support, when it does exist, is usually accessed via learning development or EAP units rather than embedded within the content-learning, assessed curriculum. Furthermore, these units seem to be housed differently in different institutions, with no clear and specific location; Davis (2019) describes the difficulties she had locating EAP managers and their Centres for her research; Hadley (2015) describes EAP practice as occupying the 'third space' in Universities. In this way, academic English 'is routinely sidelined in the institutional discourse of higher education ... in order to focus on what is considered more 'important', namely ideas or content' (Turner, 2011: 3).

However, as well as operating on the margins of academia, EAP is also operating at the heart of drives for University internationalisation.

Thus, despite the institutional sidelining of EAP units, when working to understand the role that language plays within disciplinary knowledge communication it is of great importance to consider the perspective of those who focus specifically on the development of that language as well as those whose main focus is the 'ideas or content'. This involves a questioning of the role and purpose that the teaching and learning of English for Academic Purposes (EAP) plays in the nexus of language and access to content knowledge.

Teacher or Academic?

EAP positions itself as both an academic field and a service to the rest of the academy; Ding and Bruce (2017) have described EAP, with no clear disciplinary home and uncertain identity, as vulnerable to privatisation. Ding and Campion (2016) and MacDonald (2016) have written about the lack of common title for professional roles, outlining how roles are misunderstood by others, resulting in a career structure that lacks transparency, promotion opportunities or rewards systems and suggesting that this ultimately leads to increasing casualisation of contracts and precarity of employment for practitioners. Hadley (2015), focusing on middle management in EAP, describes the tensions that exist in this space, with divergent interests around student need, academic demands, teacher resistance to institutional processes and neoliberal financial drivers. Ding has suggested that the marginalisation of the EAP practitioner within the literature on EAP and within the academy is 'both curious and to the detriment of the field' (2019: 72). Because we are not clear as to what our own identity is and there remains a lack of agreement among teachers as to who we are we also remain unclear as to what we do. Conversations on the BALEAP mailing list (2017) have shown disagreement and uncertainty as to what we should call ourselves – lecturer or teacher; more recently the editors of the Journal aligned with BALEAP – the Journal of English for Academic Purposes (JEAP) – have suggested that 'JEAP publishes articles which in some way (and with a broad interpretation) discuss English as it is used for the purposes of academic study and scholarly exchange'(Nesi & Hu, 2019). So, should they choose to engage with the literature, EAP teachers can know what this English looks and sounds like but not necessarily how they use this in practice to meet the needs of their students in the classroom nor to what extent these needs can be generalised for classroom learning and teaching practice. They remain unclear as to where language as discrete items merge with academic skills, and the extent to which specificity can be considered in reality. The knowledge base required to translate the research presented in, for example, JEAP into practice remains broad ranging and is gained, or not, via uncertain routes.

EAP has, therefore, uncertain and mixed purposes. Many of those who teach EAP, like me, have moved into it from working in the private sector where English language teaching is a business as much as an educational practice. The subject matter used in the more general language classroom is largely the vehicle for, rather than the primary concern of, teaching and learning. Although EAP attempts to distinguish itself from more general EFL/ESOL, and there is now general agreement that EAP is most effective when it is most specific in its purpose, EAP literature has rarely suggested that EAP teachers should teach content. In fact, there have been strong arguments around the difficulty of doing this (Belcher, 2006; Dudley-Evans, 1997; Morgan, 2009; Spack, 1988 in Pennycook, 1997). If content should be avoided, or at least not become an expectation within the range of expertise of an EAP practitioner, this again raises the question of what the place of EAP practice in the academy actually is. What knowledge do we develop and teach? It is my intention that this book should work to bridge these divides and gaps and suggest ways in which EAP can position itself more centrally.

The data collected from the EAP unit in my investigation clearly reflects the tensions and ambivalence of purpose engendered by the current situation. Just as with the teachers in other parts of the institution, there emerged a strong thread of collective identity amongst the EAP teachers who participated in this project. This is perhaps more surprising in this context given the lack of clear 'tribal' boundaries (Becher & Trowler, 2001) in terms of disciplinary knowledge bases, required and accepted routes of entry into the profession and general acceptance within the academy. While teachers in the other sites expressed a clear identity as researchers, or at least as research-focused teachers, EAP teachers positioned themselves most clearly as good teachers. They saw themselves as experts in intercultural communication, and in language teaching. Although not necessarily able to fully define what the 'language' they taught was or should be, this seemed to be of less importance to their sense of self than their general ability to teach, to manage a class and encourage all students within it to participate fully. This learner-centred approach has been presented in EAP literature as one way of working with students across a range of disciplines, relying (often incorrectly, particularly at TPG level) on the students' superior content knowledge as a way of working with them to support their academic development (Dudley-Evans, 1997). The extent to which EAP teachers view 'being a teacher' as vital to their work is made clear here, where one reason for student engagement with a text relating to their future discipline was seen to be the writer's identity as a teacher, as a clear communicator rather than because of the ideas expressed within the text:

- *Some of them were so excited by Stuart Hall because he's a teacher. (LC8)*

An EAP teacher's main, self-expressed, aim seemed to be around initial confidence building so that students were able to enter the academy with a better knowledge of what was to come. Similarly, insessional work was viewed as the provision of reassurance and maintenance of confidence. This chimes with the work of Wright (2017), who concludes that the socio-affective and confidence building strategies incorporated into EAP teaching are where the real learning gain occurs. This sense of purpose could also, arguably, be seen as fitting into the accommodationist views of EAP, where its purpose is seen as either simply delivering study skills to students or working to socialise them into the current norms of the academy (Benesch, 2001; Jenkins, 2013). However, given that lack of confidence is one of the issues identified by students as reducing their cultural capital, the importance of this role should not be dismissed.

There is general agreement that EAP teaching should work to enable students to access and communicate within and through the discourse used by a particular academic community. While it is generally agreed that discourses and genres are different across and within the disciplines (Elton, 2010; Kreber, 2009; Nesi & Gardner, 2012) and, therefore, EAP teaching is more effective when located within one field, it is important to acknowledge that the reality of this will entail EAP teachers needing to address questions around disciplinary content knowledge as well as language within their classroom. As I have already outlined, the EAP unit that was the focus of my study had, for the first time, moved towards what it was calling a content-based approach to pre-sessional teaching. This involved working in partnership with a disciplinary academic, who provided reading content, guidance over final assessment tasks that would familiarise students with future genre expectations and gave three lectures or practical teaching sessions during the programme. In this way it did not conform to the understanding of content-based or Content and Language Integrated Learning (CLIL) teaching as conducted in the bilingual context; nor was it intended to become English as a Medium of Instruction (EMI) teaching. It was simply EAP teaching but with a content focus that would hopefully be more familiar or of more interest to the students on the programme, preparing them better for future studies (Bond & Whong (2017) following Brinton & Holten (2001)). It did, however, require EAP teachers to work with academic texts of a discipline that they were not necessarily familiar with, and focus their teaching around a discipline that may or may not involve different pedagogical approaches. It was a move towards specificity, while pragmatically accepting that disciplinary boundaries remain a broad church, covering a range of overlapping and contradictory discourses. It was this change that challenged the EAP teachers in my sample and called their identity into question. As one teacher put it:

- *It's actually a philosophical question because is this the Kindergarten for the students to be going on to the Big School or does the Language*

Centre have a philosophy of its own and a way of teaching its students whilst they're here, regardless of what happens when they go?' (LC5)

Another recognised that the shift in approach was identifying a need to:

- *perhaps more clearly define our role in what we can and can't do and to have it clearer for ourselves. (LC2)*

The need to address content in the EAP classroom had a profound effect on all the teachers during the period of data collection. All expressed a lack of confidence in dealing with the content, acknowledging a lack of expertise, a reliance on others to provide support, and moments of stress. Observations of teaching supported these feelings, as a number of teachers demonstrated a lack of academic authority and purpose around the content.

- *It was a massive learning curve for me. Yes, of course. I found it quite stressful but I don't think at any moment I felt as though I was swimming. It's more the case that I was simply concerned that maybe I wasn't as good as I needed to be. (LC1)*

Meeting the needs of the students remained a primary concern for these teachers as they considered the possibility that they had misunderstood concepts and therefore misguided the students, but what these needs were was becoming less certain:

- *It's more with the people who are on that borderline and you're trying to get things through to them and because they're not getting it then you end up because you knew as well, you start doubting, 'I told them the wrong thing. Maybe this is all absolute nonsense', you know. (LC1)*

In this way, the EAP teachers were recognising the Threshold Concepts within the disciplinary focus and were concerned that their lack of expertise may actually be adding to rather than removing student confusion.

There was also a concern that a focus on the content detracted from the teaching of EAP. This suggests that the teaching of EAP is viewed as something separate and outwith the content, rather than flowing through it. Again, the sense of loss of academic authority preventing the teachers from fulfilling their role as they perceived it is described here:

- *I think if we really struggle with reading and understanding these texts ... because I personally was reading a text and I was really struggling with the understanding and then I couldn't spend more time on the EAP side of it. I was just trying to do my best to understand the text and that's when they asked me some questions. I'm not an expert in this, just like everyone else, but I didn't really want to have no answers to any of their questions, so I needed to understand it. That actually led me to not focusing on the EAP side. (L10)*

However, when pushed to explain what the 'EAP side' was, there remained little depth of understanding. Most teachers were unable to go beyond the following statement:

- *I just don't feel they were taught skills and language. The emphasis was wrong, so I'm not happy! (LC3)*

Again, when pushed as to 'what skills', 'what language' there was no response. This statement, expressing a sense that as EAP teachers they had not fulfilled their primary role by not teaching skills and language is thrown into sharp contrast by one teacher who questioned whether a focus on language and skills would actually meet the students' future needs:

- *Are the department bothered about skills and language? (LC6)*

Jenkins (2013) and Turner (2011) have argued that departments do not generally take language into consideration and view it as a bolt on to be dealt with elsewhere; my own data (see Chapters 5 and 6), supports the sense that disciplines would prefer not to consider language as an issue within their teaching practices. However, this does not mean that they are not 'bothered' about it, but that they have either not really considered it or they do not have the expertise, knowledge or resources to support their students in this way. It is this gap that EAP practice should be working to fill.

Despite the concern around dealing with the content amongst the EAP teachers, there was, to a large extent, agreement that building a programme around it was more likely to meet the needs of the students and provided a greater sense of purpose. Throughout the summer period, I observed both teachers and students who were fully engaged in the work they were doing. Teachers commented frequently on the high levels of engagement across the cohort, making statements like 'they're a higher level of student this year'; 'absolutely everyone has done their homework!'; 'they were all talking'; 'I can't believe how much work they did'; 'they can really see the point of it all; they can see how it's going to help them'. They were able to contrast this to previous teaching they had done:

- *I think at the very least though we have a duty of care to students to give them an insight into what a Masters is, and doing a little lesson on cohesion where we match a few sentences ... they don't need to come here to do that. ... So, I think fundamentally we were doing more ethically the right thing. (LC1)*

It was not only in the students that the teachers on this programme observed a sense of transformation. Some also expressed a similar shift in understanding and engagement in themselves, as this email sent to me at the end of the teaching period suggests:

- *It's the first time (and this is my sixth pre-sessional) that I have seen the transition between pre-sessional and Masters course. It made all the hard work seem really worthwhile. I also seem to have had a kind of conversion about the content. Having been resistant to it at first (I blame Bourdieu), I found the work on identity absolutely fascinating and also personally relevant. You might think that I would want to switch off when the course finished, but actually I had a long discussion with my son about it all and ended up recommending that he read Bourdieu! (LC6)*

Having spent some time grappling to understand the subject knowledge, the EAP teachers thus began to see how they could reassert their 'EAP selves' on the content, and felt that they would be able to be clearer in their purpose when asked to teach the programme again:

LC11: *Yeah, to teach them how to deal with these articles.*

LC8: *To teach them what to look for. What is an important sentence in this page and what is less important?*

LC5: *I think that's right. I think the other thing is that I really think we should make it far more obvious how difficult we've been finding this because they need some roles. They need modelling.*

LC6: *There just isn't a separation, is there?*

leading them to the conclusion that it was not really possible to separate the content from the language and discourse used, but that it is key to strike the correct balance:

- *... the question is how far should we, say, be teaching skills like discourse and how far should we be teaching things like the structure of a society, what power is and things like that because we're not subject specialists.*
- *we're so focused on the content and that's what they were having problems with. Like with the BBC thing they were just so narrow in their view of what that question might mean, you know.*

In this way, then, the EAP teachers started to argue for greater integration between content, skills and language as a way of enabling them to fulfil their perceived roles and to provide more holistic guidance to their students.

For the insessional teachers, their total separation from the content teaching was seen as detrimental to the students' learning. EAP teachers felt that they were teaching in a vacuum and were not able to guide students to apply what they were teaching in a useful manner:

- *For instance, you watched me do a summarising and paraphrasing lesson. I think there's more room, at the moment because that was an abstract concept taken away from what the students were doing at that particular time. (LC9)*

What emerges from this data then is that EAP teacher identity in this context was based very clearly around a sense of being a 'good' teacher and good intercultural communicator. This expertise is measured by teaching-focused qualifications and experience, rather than academic credentials. It is also measured in contrast to 'academics' in the institute who do not have similar qualifications and do not focus entirely on teaching. However, it is also closely connected to the isolated teaching of (general academic) language and skills, where the teacher and students work closely together in a protected environment where a key focus is building confidence rather than developing academic authority (of either teacher or learner). When required to take this 'good teaching' outside the confines of a general EAP classroom and work in collaboration with disciplinary content, the EAP teacher identity is brought into question. Language and skills can no longer be viewed as a teaching purpose in and of themselves but are more clearly bound up in cultural and contextual practices. This, in turn, calls the identity, knowledge base and choices made in EAP practice into question. Being a good teacher is no longer an adequate identity, as it becomes important to ask the question: a teacher of what? And based on what? And who makes the decision around what should be taught? In this way, agency becomes entwined with identity.

(Denial of) Agency

One of the central tenets of the EAP field is that curricula, syllabi and all teaching should be based around meeting the needs of the student (BALEAP, 2014; Bruce, 2011; Ding & Bruce, 2017). This is in contrast to most traditional UK HE syllabi, which is centred around an accepted canon of knowledge within a particular field. This needs-driven agenda is viewed by EAP practitioners as one of the strengths of EAP. These needs should take into account both the entry point and exit expectations of students and on students in terms of language proficiency and academic attributes, as well as the expectations of the receiving academic departments. Practitioners generally aim to be reactive to student feedback, to meet the needs of the students they are working with and are keen to engage with dialogue and learn from disciplinary academics. This, it could be argued, is where the identity as a 'good teacher' stems from. EAP has a 'student as partners' focus (see, for example, Healey *et al.*, 2014); however, this focus, without clarity around the academic knowledge base on which to build a partnership, could also lead to directionless teaching and learning with no clear specific outcome.

EAP is, also, a highly collaborative endeavour. EAP practitioners are rarely required to make curriculum decisions alone. Beyond a sense that the needs of academic departments and of students dictate choices around what should be taught in a programme, EAP teachers can often feel that they do not have the opportunity to act as they would like because they are also controlled by less benign forces that are outside of their control. These (neoliberal) forces can include governmental visa regulation changes; other Schools' international student admissions policies; management decisions within the EAP teaching unit; marketing policies and the choices made by those responsible for timetabling and staffing allocation.

Without a firm sense of academic identity or of the value of EAP teaching in its own right, if you are always reacting to and building work around the needs of others and reacting to external policy decisions, it is easy to lose a sense of agency. A lack of agency can, in turn, have a direct impact on the work we do and our understanding of the purpose of our teaching.

Unlike most teaching in HE, at least within this institution, the programme that an EAP teacher is likely to be teaching on can be uncertain up to a matter of days before the programme begins and can change from term to term. During summer pre-sessional teaching, there is also a large increase in the numbers of teachers working in any one EAP unit. This had real practical consequences for the teachers involved, and seemed to remove a sense of stability and control over the working environment:

- *I suppose on a practical level actually it's not necessarily the course but it's things like moving to a different office and people wanting to chat all the time and actually I just want to get on with my work 'cause I've got … you can't just sit at your desk and read what you need to read and concentrate on it properly because somebody wants to start talking to you. (LC1)*

A further impact of the ever-changing teaching allocations was expressed as a lack of opportunity to engage with the content material of any one programme both before teaching begins and, for some, during the programme itself:

- *[you] could have passed it over to me to say you've found an article you thought might have been interesting, but then you wouldn't have known that I was going to be on the course so I think it's all the sort of practical things like that. If I had known even a month before I could have like taken a journal article home a week ... (LC1)*
- *I felt very disconnected from the content because I've not been involved in the content side at all. (LC4)*

By not having defined areas of discipline specialism, teachers felt at a further disadvantage when working with students through content and were unable to take full control of the language teaching they did. This was highlighted in observations, where teachers struggled to demonstrate academic authority over the content and were therefore unable to exploit this fully for language learning purposes. There was over reliance on materials provided for them, without agential teaching decisions made around them. Teachers were constrained by the syllabus and teaching materials, which could be seen as being imposed upon them and which they felt unable to question:

- *you're on this conveyor belt. You're sort of going through, going through, going through and just, you know, throw bits of paper at people. Well you might as well just throw bits of paper and then walk out the room, you know. You're not teaching really are you, you know. (LC1)*

There also appeared to be a lack of clear EAP disciplinary agency; a lack of understanding that EAP has, or should have, its own principles and outcomes; and that EAP teachers can exert their own knowledge and understanding on students' development. The future academic School, and their needs, purpose and assessment were privileged by EAP teachers over their own beliefs and knowledge around language, literacy and discourse. Thus:

- *we get a feel for what is a reasonable essay in their [the School's] eyes. (LC2).*

EAP teachers were, for example, unsure of how much their own assessment of student work had relevance to decisions around their suitability for study on an academic programme, something that had a direct impact on Mai's trajectory onto her TPG programme. In Bernstein's terms (2000), EAP forms the interface between knowledge production and practice. EAP teachers seemed to be positioned, or position themselves, as the 'outsider within'; the purpose of EAP was couched in terms of meeting the demands of others, rather than following its own agenda:

- *The other stakeholders are important in this, aren't they? If the students are all completely happy and the [School] is completely happy ... then we don't really need to worry about it. (LC4)*

This 'not needing to worry' ultimately reduced the need to take responsibility for possible outcomes. The conflict around not feeling ultimately responsible, yet being deeply concerned, for students and their confidence and well-being continued into the high stakes area of assessment. The comments and observations again reveal a lack of clarity, purpose and agreement, touching again on issues of need, confidence and the separation of language, skills and content. Working with an assessment question that required engagement with the subject area through spoken and written texts moved some away from their comfort zone:

- *the level of difficulty of the questions because with my group, I spent a long time for us trying to get our heads round what the questions meant and the content was so heavy, I think the skills and the language got a bit lost for us. (LC3)*

Others felt that they were able to consciously separate the two aspects when grading an essay,

- *We could theoretically give a good mark to something that's fundamentally wrong if it's well structured and fundamentally inaccurate because we can't know everything for every subject. I'm not shutting that part of my brain down, but it's definitely the other part that's more active. (LC9)*

whilst some accepted the intangible and subjective nature of assessment practices and relied on their teacherly experience to allow them to place students within a band:

- *if you're an experienced marker, you kind of know anyway. You know realistically that that overall is going to be an essay which would pass and that one is a bit too weak somehow but it's being able to justify it isn't it. (LC1)*

For some, the whole purpose of assessment also came under question. Should there be any value placed on the end product of EAP teaching if the main role is to prepare students for future studies?

- *The output at the end; is it the process or is it the result? (LC5)*

This links to the sense that the role of the EAP teacher is to promote and raise self-confidence and highlight progress through a long learning journey. Teachers noted when students recognised this and used it as a measure of their own success:

- *They were actually comparing their essay results with their summary … [saying] I'm really happy, even though my mark is in the 50s band but I'm really happy. (LC10)*

One teacher, in fact, believed that it should not be the responsibility of the EAP teacher to assess students at all if the assessment was to include any focus on the content knowledge, and that the EAP teachers' assessments did not matter unless they could be shown to clearly tally with those of the content teachers. This was in spite of the Learning Outcome for the pre-sessional programme, and the assessment criteria, having a clear EAP, language, genre and discourse focus:

- *I don't think it matters what we would give them. I think it matters what they would give them and that's why I was uncomfortable marking. (LC3)*

Much of this belief seems to stem from a lack of confidence on the part of the EAP teacher in dealing with the nuances of academic knowledge communication. Beyond the basics of language as a code; building a formulaic essay through the moves outlined in, for example, Swales (1990), the work becomes more ephemeral and specialist, and therefore difficult to pinpoint.

- *we were given a range of essays [by the School] but I decided not to use them and the reason being was the reason the essays got lower marks was on deep points of content. They really weren't that visible to a non-expert. It wasn't like the paraphrasing's weak and that's why they got it or these sorts of tangible things that we could say that's why that got the mark. It was that the overall argument was not … I could see that the one that got 80 was much better but I couldn't pinpoint what was wrong with the argument because I don't have the depth of subject knowledge. (LC9)*

Any move away from the more remedial/deficit approach to language teaching seemed to become problematic. This chimes with the argument that the transformative agendas of academic literacies and critical EAP will be rendered futile if 'there are not practitioners in place with the agency and capital to enact, develop and critique theories, ideas, research and ideological projects' (Ding & Bruce, 2017: 206). There was a sense through all of the interviews with the EAP teachers, therefore, that they felt they knew what the students needed, but they were not entirely sure how to break this knowledge down to meet that need. In fact, there was no clear general consensus over where the needs were greater and what should be given priority; few were able to clearly define what EAP was, or what language should be taught as a priority. Full agreement was really only achieved over the need to boost students' self-confidence and provide friendly, occasionally emotional, support as the student transitioned onto their academic programme.

It seems then that, in part due to working conditions that often require delivery of prescribed materials, but also because of an abdication of agency, EAP teachers often chose *not* to make purposeful, knowledge-based decisions about what they teach. They are not really clear what the knowledge base they should be working from is. This resulted, as Brooke (2019) suggests, in 'academic language courses [that] tend to be devoid of a theoretical approach to education but places the teacher in the role of linguistic expert'. If EAP teaching is based on linguistic expertise, but is unclear as to where the linguistic focus of a programme or learning episode should be because responsibility is placed within the discipline; if EAP is simply catering to the demands of others, then its teachers do not have to think too deeply about their own purposes. They are able to abdicate responsibility for decisions around curriculum, content and what measurable learning gain may or may not occur. EAP teaching with this abdication of agency perpetuates both the suggestion that EAP adopts 'an unquestioning stance towards the departments and disciplinary practices that students encounter' (Benesch, 2001: x), and that EAP follows an accommodationist view of education with a remedial positioning of language work which, according to Turner is necessary 'to maintain the culturally embedded and socially embodied "habitus" of being academic' (2011: 37). In this way, EAP teaching becomes a pharmakon, a scapegoat for when students are failing to communicate, a poison in perpetuating myths around language as being easily fixed and isolated from academic thought processes and only a remedy as a placebo, providing confidence and emotional 'sticking plasters' to students without treating the much more complex cause of their difficulties.

Developing Confidence

It is possible to draw a number of thematic parallels across the data I gathered from the EAP teacher participants and the content teacher participants. For both groups, identity, agency and temporality were clear areas of concern. However, for the content teachers, I have already outlined how a sense of identity and agency was drawn and developed mainly in relation to their academic, researcher selves and was measured against peers and colleagues within their discipline; any loss was perceived as emerging from institutional neoliberal activity with changes in student behaviour and responses to teaching being a part of this. Students were largely viewed collectively, *en masse*, as one part of a larger academic endeavour. This contrasts with the emerging patterns around EAP teacher identity, which stemmed almost entirely from interactions with students as individuals.

This difference became clearer when considering the emerging themes of trust or confidence. While the same kind of language was used by both groups, the focus of the emotion being described was different. For the content teachers, *trust* was placed in colleagues and can be conceptualised as significant networks (as described in Chapter 4). EAP teachers rarely commented on the work of colleagues; EAP pre-sessional teaching is by necessity a collaborative endeavour and these networks are a taken-for-granted element of this, so generally unworthy of comment. Instead, language around gaining trust and developing confidence was aimed very specifically at the role they played in relation to their students. This work can be conceptualised as developing *emotional capital*. Teachers spoke of their work as being around providing support, having patience, giving up time and developing close relationships with their students and felt that success in this area was worthy of comment:

- *I mean I was, I actually got to know the students, you know. (LC1)*

Watson Todd, suggesting there was a need to redress the balance of focus away from the content towards the methodology, or the 'how' of EAP teaching concludes that as 'teachers we need to remember the students' learning needs as well as their language needs' (2003: 154). The data collected suggests that meeting the needs of the students *is* central to the work of EAP teachers and core to their sense of self. Teachers had quite strong emotional reactions when they felt they were not able to meet or recognise these needs:

- *I just thought, 'I want to know what these students are thinking and how they're thinking' and I couldn't do that because they were reading somewhere else and I was doing the kind of writing. That was amazingly frustrating for me'. (LC5)*

Working to meet the needs of a particular group of students requires teachers to react to context, task and time-specific situations. This makes planning a coherent programme problematic and, ideally, requires teachers to spend time getting to know each student and work with them on an individual basis. It is in this area of work that it is possible to see more agential decisions being made. Teachers reported attempts to address individual requirements regardless of programme or time constraints:

- *I think there needs to be space for teachers to monitor and be able to address concerns that students have. (LC1)*
- *We never stuck to 15 minutes [consultation time] because well it's like you either help them or you don't. What's the point in having a consultation if you go in and just go, 'Oh, you spelt that wrong. Bye'. (LC1)*

Although some of the focus was on their academic needs, much of the concern was for the students' welfare and building up their general levels of confidence:

- *I think [it] put a lot of pressure on the students. (LC7)*
- *a student said to me they really enjoyed coming to the Insessional classes because it was a break from what they were doing on their Masters ... where they can be free and they're not going to be judged the same, especially at the beginning. ... At the end, they really started to come on and they were much more fluent and* ***confident****. (LC9)*

It seems that students valued this focus that EAP teachers placed on being friendly and accessible and highlighted the value they gained from this confidence-building approach as much, if not more, than any language or skills they may have developed through an EAP programme. Students reported feeling supported and welcomed, and entering their department with a sense of optimism and confidence, as these comments from the Student Satisfaction Survey suggest:

- *The programme has helped me to present myself with confidence.*
- *I think the tutors are good and they help me a lot. When I have some questions, they try their best to help me to solve.*
- *More confidence than before. Enjoying the time we works together in a group.*
- *Tutors are friendly to everyone, we can learn something in a comfortable circumstance.*
- *tutors were very kind to help students deal with their language learning problems tutors tried to encourage some students shy to express themselves. Students got along well with each other, and they could work together happily.*
- *All of my teachers are very nice and patient!*

Those students who were interviewed once they had moved into their School also highlighted the importance of this confidence building, emotional capital work. This came, not only from the contact with EAP teachers, but also through the development of friendships and social networks that were created through the pre-sessional programme:

- *I didn't meet yet friends from my country but at the beginning, I was afraid that I would be misunderstood when I speak because sometimes, the language does not help me to deliver what you want to say. I found that it's very interesting to have friends from different cultures and I enjoy them. I'm learning from them a lot. Yeah, we're becoming now closer, I believe. I think that it makes me also more confident that it's okay to make mistakes and it's okay to just practise and it will be better with time. (5B)*

This focus on the individual, emotional need was an aspect of their work that EAP teachers felt skilled in, and confident that they did well. While teachers also voiced quite strong opinions about the academic needs of the students they were teaching, they did so with less certainty and clarity, as is demonstrated by the heavy use of hedging (see bold script) in the quotes below. Teachers also seemed to consider what students wanted, or expected (as consumers of education and customers paying for a service), to be of equal importance as what they needed:

- *what I was originally going to say was simply the fact that it depends,* ***to some extent****, on the students, in as far as the content goes, what we're actually teaching.* ***In the broadest terms, I think it probably is*** *skills, but* ***then it depends specifically*** *according to the class. I've brought a lot more speaking into the lessons because I noticed that that's what they needed. In the group, it might be actually some grammar* ***because that's what they really want****. (LC1)*
- ***I think they kind of expected help*** *with that and how to write academic essays and the conventions. (LC7)*
- ***my feeling is*** *that there was too much reading. I think what you're doing is you're assuming that we're going to have the same quantity for next year but my feedback would be there was too much and the students, I think at certain points, I think felt they were drowning in it. I would like to see it pared back a little bit; not particularly for our sake, but also for us, but more for the students really. (LC3)*

This final quote brings together and also highlights the conflict between the emotional care of the EAP teacher (*I think felt they were drowning in it*) and concern about meeting the academic needs (*there was too much reading*). This is echoed by this student:

- *Sometimes I suffered from a lot pressure.*

However, LC3 also highlights one of the other broad areas emerging from the data from EAP teachers: that of dealing well with the disciplinary academic content themselves, or not wanting to do so (*I would like to see it pared back a little bit; not particularly for our sake, but also for us*). There seem to be two key reasons for this. The first relates to a lack of EAP tutor knowledge – both of the disciplinary content, but also of core EAP knowledge in terms of working to decode this knowledge – in fact Watson Todd's argument (2003) that there should be a re-focusing on 'how' is superseded with the need to consider the 'what' of EAP. The second links, again, to temporality; to the amount of time available to read, decode and work with students on a text in a meaningful manner.

Time Limitations

Time, or rather lack of it, is a theme that threads through all of the data collected in relation to the teaching and learning of EAP across all three Case Study sites. This lack of time led to a questioning of the effectiveness of any of the teaching that took place, and of what the core function of that teaching should be. This questioning was also echoed by teachers from other Case Study sites, who also wondered to what extent separate language-focused teaching could have an impact on students' ability to access TPG content:

- *there is a point beyond which no amount of resourcing is going to help 'cause it takes time and practice to learn a language and to use that language effectively and if the student comes in, you know, with language skills that, you know, I have real doubts about what the different IELTS scores actually mean in practice. (M4)*

As Turner (2004: 97) has pointed out, students 'seem not to want to spend time, effort and money on getting to grips with the language, but to proceed as quickly as possible to the 'real' thing'. Within the EAP unit, teachers expressed a sense of resignation over their inability to cover enough ground during the six weeks of a summer pre-sessional:

- *I mean in six weeks you just couldn't cover everything that they needed to know but yeah, I think that what we did was a very good start. (LC1)*

This led to a general questioning of the purpose of the teaching:

- *You've got six weeks and you've got to prioritise what are the most important things. Are we supposed to be struggling through the content with them more than we're supposed to be teaching them the skills to get them on the road when they're there? (LC3)*

EAP has generally been complicit in, and profited from, the myth that it is possible for students to improve their IELTS grade by, on average, 0.5 over a 10-week pre-sessional programme. However, many pre-sessional programmes that are delivered by UK university EAP centres do not explicitly teach or test through IELTS, and claim to be helping 'learners gain access to ways of communicating that have accrued cultural capital in particular communities, demystifying academic discourses to provide learners with control over the resources that might enhance their career opportunities' (Hyland, 2016: 6). All in six to twelve weeks! It is hardly surprising that there was a frequent expression of frustration at not having the time to work with the students in great enough depth, of never quite getting to the crux of the matter because of the need to pack in too much information beforehand.

Teachers also pointed out the lack of time available to them for preparation, not specifically of classes, but in terms of their own knowledge building around a discipline. They argued that not having time to read and absorb key texts they were working with, not being able to attend seminars in Schools that would provide an insight into the subject they were working within, prevented them from meeting the desired learning outcomes with their students.

- *[If there had been time to prepare before] I would have been able to digest it and feel much more confident rather than just being like half a millimetre ahead of the students sort of thing, you know, and relying on the lectures as much as they did for clarity. (LC1)*

Time is also an issue for students' EAP development. Once they have begun their Masters programme, the study of language is no longer a priority. Students with the lowest levels of language proficiency are likely to need to spend more time reading and understanding the content of lectures and texts as well as ensuring the quality of their own writing. They need to prioritise the work they do for this. Therefore, despite having the greatest need for further focus on language development, they frequently chose not to attend language classes due to lack of time available to them. This brings into question whether the core EAP purpose of working with students to develop language to support their academic studies is being met through offering adjunct classes separated from the content.

The Knowledge Base of EAP

At this point we return to the distinction between an EAP practitioner as a 'teacher' rather than an 'academic', and within this consider the extent of the knowledge base that it is necessary for the practitioner to draw on. The data that speaks to the themes of temporality, confidence, identity and agency also suggests that the core element that is missing is a strong foundation of EAP knowledge.

There is no agreed EAP career pathway via distinct, required professional or academic qualifications. BALEAP, the professional body for EAP in the United Kingdom and, increasingly, globally claims that a core aim of its teaching EAP competency framework is to 'enhance the quality of the student academic experience through facilitating the education, training, scholarship and professional development of those in the sector' (BALEAP, 2014: 4). However, this facilitation is only available to those who are already involved in the practice of EAP, who are able to build and provide evidence of their practice in action. Entry into the profession remains largely based on language teaching qualifications – primarily the Certificate and Diploma in English Language Teaching (CELTA and DELTA respectively; the former of which can be gained from 120 hours of study) – that meet the accreditation requirements of the British Council, provide some evidence that an individual can teach (i.e. manage a classroom; ensure communication takes place; plan and teach discrete language items and skills). Reluctance to shift these basic requirements is based, I would suggest, on the continued need to recruit large numbers of teachers to work on short contracts over the summer in order to meet the exponential growth in students taking pre-sessional programmes. Any EAP teacher who wishes to develop a career in EAP needs, as Campion argues (2016), to accept not only that 'the learning never ends' but that there is no clear direction to the learning that is necessary and that it may need to be undertaken with little professional guidance.

The current knowledge bases for EAP, as touched on in Chapter 1, focus largely on written communication and on identifying linguistic and discursive norms in academic communication. Yet, as many (Turner, 2004; Wingate & Tribble, 2012; Jenkins, 2014; Hyland, 2016; Lillis & Tuck, 2016) have argued, there is actually little agreement as to what these norms actually are or in fact whether the work of EAP is to help students to replicate these norms or to transform them. To add further complexity, EAP practitioners work across a range of contexts (monolingual; multilingual; English as a Medium of Instruction; interdisciplinary; transnational, for example) and, if not controlled and confined by a fixed programme and set of materials, are able to draw on multiple theories of language and tools for learning. The extent of the disagreement over what EAP is and what should legitimately be included in an EAP curriculum is so great that it has led Monbec to describe the teaching of EAP as a 'field of struggles' (2018: 92).

The context, conceptualisation and knowledge base of EAP practice is, therefore, heterogenous. Further confusion is created by the frequent conflation of pedagogy, curriculum and knowledge (Kirk, 2018). These three aspects of EAP work should be viewed as different fields of knowledge, of reproduction, recontextualisation and production respectively (Bernstein, 2000). The majority of EAP teachers are primarily involved in the 'reproduction' of knowledge as they engage with various pedagogical practices in their classroom. In order to do this, they need to be able to interpret the intentions of the curriculum as it works to recontextualise knowledge production. It is within curriculum design and planning – the bridging space between knowledge production and reproduction – that the complexity of EAP practice is most evident. The content – the carrier of language and discourse – used within EAP curriculum planning does not reflect the knowledge produced by EAP research; rather EAP curriculum designers need to interpret the EAP research and use analytical tools to de-code subject content and guide students (and other EAP teachers) through the process so that they are able to reproduce the knowledge gained across different contexts. In order to do this, EAP research suggests various approaches, including Systemic Functional Linguistics; Genre Theory; Discourse Analysis; Corpus Linguistics; Academic Literacies and, more recently, Legitimation Code Theory. It remains the work of the practitioner to translate these tools for knowledge production into tools for recontextualisation and finally reproduction. Ian Bruce (2005, 2008) in particular has suggested a set of analytical tools that can be employed for EGAP writing syllabus design based on cognitive genres and rhetorical types. However, this is a complex process and one which few practitioners have the knowledge, training, confidence or time to enact. Therefore, the practice of 'EAP appears to lack the tools to see and analyse the forms that these knowledge practices may take across different fields of practice' (Kirk, 2018: 64).

Beyond the knowledge base and need for professional development within the field of EAP, it is also necessary to acknowledge that '[s]tudents do not learn in a cultural vacuum but are judged on their use of discourses that insiders are likely to find effective and persuasive' (Hyland, 2016: 19). Increasingly, the EGAP/ESAP (English for general or for specific academic purposes) debate seems have been theoretically resolved. The continuing focus on EGAP is largely due to financial and logistical constraints rather than any evidence-based suggestion that it is a more effective way for students to learn. As practitioners begin to move, then, towards a more ESAP focused approach, it is also necessary to consider what, and to what extent, they might need to have an awareness of the content knowledge of the disciplines.

Content knowledge in EAP can be viewed from different perspectives. Teachers from the EAP unit in my study separated the academic content, which they denied having expertise in, and 'language and skills' which they claimed expertise in. Thus, academic skills and linguistic content, including linguistic terminology to describe the language system and codes used were viewed as EAP content knowledge. There was much evidence of this language being employed in observed classes, based on a teacher assumption of a shared knowledge with students who were long-term language learners. However, this 'essentialist view of language in which language is understood to be a decontextualised skill that can be taught in isolation from the production of meaning and that must be in place in order to undertake intellectual work' (Zamel, 1998: 253) devalues the intellectual work involved in learning and working within a foreign language, and suggests that language is simply a matter of training prior to the real work, rather than an ongoing enterprise developed through interaction with the subject matter.

It was this view of language as a separate, transparent vehicle that teachers seemed to employ when it came to assessments in particular, suggesting that it was only possible, and was in fact necessary, for an EAP practitioner to consider linguistic proficiency in a content vacuum:

- *We could theoretically give a good mark to something that's fundamentally wrong if it's well structured and fundamentally inaccurate because we can't know everything for every subject. I'm not shutting that part of my brain down, but it's definitely the other part that's more active. (LC9)*

In some cases, this went even further and returned to the idea of being an 'expert teacher' rather than a teacher with a clear knowledge base and sense of the specific learning outcomes being measured:

- *if you're an experienced marker, you kind of know anyway. You know realistically that that overall is going to be an essay which would pass and that one is a bit too weak somehow but it's being able to justify it isn't it. (LC1)*

The second view of EAP content knowledge, taken by some participants, was that of the disciplinary content through which an EAP teacher and students were working. Again, there was much evidence of this being worked through in observations, but with less confidence. It should not be the work of the EAP teacher to 'teach' this content; this was made clear to teachers who worked on the pre-sessional programme and all content lectures were provided by experts working in the home School of the students. However, lacking confidence in how to separate the content work from the EAP work, and not being clear how to focus student attention onto the latter resulted in a loss of confidence and a decreased sense of authority:

- *I think the order in which things appeared ... although logical in some sense was very, very difficult for me, as a non-expert. (LC5)*

Teachers also expressed a concern that the inclusion of content to the curriculum meant that the students who were most in need of language support were missing out on learning opportunities, particularly around language development. Again, they connected this to their own lack of knowledge, expertise and confidence in teaching discipline specific EAP

- *the level of difficulty of the questions because with my group, I spent a long time for us trying to get our heads round what the questions meant and the content was so heavy, I think the skills and the language got a bit lost for us. (LC3)*

The change in focus to a content-driven approach led to awareness of a mismatch between the 'language acquisition' kind of teaching and the intended learning, as encapsulated by this teacher:

- *the majority of the input is study skills based, whereas the student output is more literacies based. So the bibliography task and the seminar task are probably more literacies based, but a lot of our input is kind of study skills based. (LC9)*

This was partly seen to be due to a disconnect between the teaching of academic language and skills and the subject specialism of the students. In order to move beyond the generic teaching of language and skills, there needed to be a clearer link between the two while maintaining a focus on language, discourse and metadiscoursal features. There was some resistance to this change in approach, resulting in levels of divergence amongst EAP teachers. Thus, some remained focused on more generic understandings:

> **LC3:** *I think we should have been guiding them much more in class with their actual essays and really going through it properly to develop their skills.*
>
> **LC8:** *Yeah, even just language for linking, for example.*
>
> **LC3:** *Yeah, I think various people did various things. For example, that was one of the things I concentrated on and I saw it come back in the essays but there were other things that I didn't have time to do that therefore didn't come back to me. I would just like time to do all those things properly, so that I feel like we have geared them up ready for their essay writing.*

while others began questioning their understanding of this skills and language approach to teaching:

- *I don't necessarily think it's for us to teach the vocab. I think it's for us to train the students to find the vocab. (LC9)*
- *[what is missing is] how to make use of complex texts to make sense of complex meanings. (LC13)*

Therefore, a third perspective of the required knowledge base of EAP/content teaching is one which connects the two together, where linguistic and content knowledge are harnessed to highlight the metadiscoursal features and forms of knowledge communication employed by a particular discipline and genre. Although this had been the intention behind the development of the programme as a whole, it was not identified as such by some of the teachers:

- *there could have been more integration of the content and the skills and that you could have taught the skills through the content, rather than having bits separate. I think that would have helped all round'. (LC8)*

EAP teachers themselves experienced a move through liminal spaces as they struggled with content, aware that their teaching was possibly too focused on disciplinary knowledge (Elton, 2010) until there was a reconciliation between working with content but highlighting linguistic features that came about as a result of greater comfort with all forms of content knowledge:

- *I got students to highlight useful phrases. Highlight what they thought was subject-specific vocabulary. What they thought was generally useful language that could be recycled into any piece of academic work. When we had a text to engage with like that, I generally tried to get students to do a bit of that. I think that's something we could probably build on more. (LC9)*
- *I'm becoming more aware of some of the very different literacies which are involved. (LC2)*

In order to do this effectively, teachers require an understanding of the content and pedagogy of EAP and, to a lesser degree, of the students' future discipline. EAP practitioners also need to employ knowledge of their own particular academic context so they can develop an understanding of disciplinary and contextual norms in order to be able to support their students' developing understanding around how to enter their own community of practice. I will discuss in greater detail what this 'pedagogical content knowledge' might look like in Chapter 8, but the general idea is exemplified below. EAP teachers working with students already on their TPG programme reported:

- *when I asked the students' questions or I gave the students opportunities to ask me questions I felt that most of the questions were linked to academic socialisation rather than the language. I had very few questions related to kind of sentence structure or paragraph structure or basic language. There were a lot of questions about plagiarism that*

I had. A lot of questions about, 'Is this okay? Is that okay?' this sort of thing. (LC9)

Thus, the confidence building element of EAP teaching that seems to form one of the main threads of agreed identity for EAP teachers was seen as key to encouraging the establishment of EAL speakers' cultural as well as emotional capital. Wright (2017) argues that this can be combined with language teaching through a joint focus on linguistics (specifically Second Language Acquisition), discourse and interactional competencies; she labels this 'Academic Interactional Competence', expressed slightly more prosaically here:

- *It then comes back down to the question again of … what's the message that we want to give to the students? That it's about being able to speak and engage. (LC4)*
- *For them, it's the pressure of time and it's this worry about how long it will take them to go from the article, read it, understand it, identify what they can use for their purpose, and then put it into their own words. That's what they worry about. It's partly a combination of good practice and just giving them the confidence to do that. (LC12)*

However, this still does not really specify what knowledge an EAP curriculum, even one that is focused on English for *Specific* Academic Purposes, should include. Much of the issue here seems to lie in the distinction, and clash, between what an EAP teacher needs to know in order to provide effective EAP-focused learning opportunities and the knowledge that our students need to access. Unlike other disciplinary content teachers, the two knowledge bases are not the same. This lack of certainty as to what EAP content knowledge is, and the continuing focus on the affective impact of EAP teaching rather than a focus on the process of building knowledge has lead EAP practice into 'knowledge blindness' where

> what that knowledge is, its forms and its effects, are not part of the analysis. Instead, knowledge is treated as having no inner structures with properties, powers and tendencies of their own, as if all forms of knowledge are identical, homogeneous and neutral. (Maton, 2014: 2)

Chapter Summary and Practical Lessons Learned: A Re-positioning

The conundrum of the position of the EAP profession, as outlined in Ding and Bruce (2017), is clear within the data presented in this chapter. Whilst asserting a clear identity as expert, student-centred teachers, who work to build confidence in their students, there was a lack of clarity or agreement as to what this should look like. Many very strongly felt that their role was to teach 'language and skills'; conforming to the 'butler stance' proposed by Raimes (1991), 'a technician, who is able to execute pedagogic technique competently...[with] no need to consider theory or research as a basis for practice' (Ding & Bruce, 2017: 9).

However, some participants expressed a gradual shift in perception as they were required to engage with the content of a particular discipline and work with this, rather than with materials written specifically for an EAP classroom. There was a realisation of the complexity of their students' needs and, through interaction with disciplinary colleagues, a vague understanding that discourse competence was assumed but not explicitly taught. This caused some to want to engage and understand further but also to feel that they did not have the time, resource or authority to do so. Others retreated to a more hygienic position, reasserting their technical role and abdicating agency altogether.

Difficulty was experienced when EAP teachers were required to move away from teaching language as discrete items and there was a clear lack of knowledge around how to de-code complex texts and discourses to enable their students to access the necessary knowledge.

Time was also an issue. The understanding of language as external to other knowledge perpetuates the myth that it is possible to move all students in a linear fashion from one proficiency level to another in a small space of time. Teachers felt the pressure to do this. As the curriculum they were teaching required something other, i.e. a consideration of language in disciplinary context, they also felt pressured by a lack of time to develop an understanding of the content and of how they might work to de-code this. Within the intensive summer pre-sessional teaching period it is necessary that the curriculum itself enables teachers to do this (see Kirk, 2018).

Unlike previous chapters, I can see no 'quick wins' to the necessary re-positioning of the EAP practitioner so that the 'teacher or academic' distinction is challenged. This need to engage in scholarship and develop an identity as an academic scholar within the academy is becoming increasingly established (Ding, 2019; Ding & Bruce, 2017). However, for teachers who only work in EAP for short periods of time (typically over the summer), this is an ambition that only a few will be able to achieve.

Those who are in more privileged positions, with more permanent roles need to do the work for our colleagues. It is necessary to ask ourselves some basic questions:

What do we need to know? How do we need to work?

I suggest that to answer these questions well, we need to focus on three areas of work:

- Ethnographic research of the academy

In this, it is necessary to focus on a specific discipline and its knowledge, to explore the concerns of academics within a particular context. We need to develop a 'nuanced and careful approach ... that can account for the knowledges that have shaped educators' identities and agency, and that are, in turn, playing a significant role in shaping students' identities and agency' (Clarence, 2016: 125). We then need to develop ways of helping academics to see how language connects and intertwines with their knowledge base and to ask questions that allow them to see this tacit understanding as new (Freeman, 1991).

- Develop a knowledge/language toolkit

The knowledge base required to translate the research into how English is used in academic contexts as well as the ethnographic research outlined above into the classroom remains by necessity broad and varied. A 'good enough' knowledge of the range, rather than a deep focus on one area, allows a toolkit approach where teachers draw on multiple ways of de-coding complex discourses. This will allow students to also view the work of de-coding as necessarily complex. In this way we shift EAP teaching, as well as the understanding of our students and disciplinary colleagues, away from the view of language as hygienic, transparent and separate.

- Create a curriculum that extends teacher knowledge as they enact it

It is through engagement with the curriculum, through the enactment of it, that the knowledge base of all EAP teachers can be extended beyond that of the technician. EAP pre-sessional teaching is likely to always be a collaborative enterprise, with multiple teachers working on the same programme and through the same material. It is through this curriculum that many will first encounter and begin to understand what EAP is. Therefore, the curriculum should work to encourage practitioner learning through the inclusion of the variety of means to interpret and de-code knowledge while concurrently meeting the needs of the students it is primarily written for.

7 Language Across the Curriculum

Simply having a diverse student body does not mean the education or even the campus is global in nature'
(British Council, 2014: 4)

In this chapter I assert that in order to achieve the aim of creating a truly internationalised University, language needs to be foregrounded across all teaching and learning in Higher Education, and throughout the extended curriculum. If we do not become aware, and raise awareness, of the importance of language and work towards an intercultural model of communication and learning, it is not possible to claim that Higher Education is either international or inclusive. 'English can no longer be cast aside in the internationalization literature as though it was merely a practical problem to be "fixed" in EAP units' (Jenkins, 2013: 11). This does not, and should not, mean that EAP units become irrelevant, but that the provision of English for Academic Purposes should become a key element of any University curriculum. I argue, then, that language should be embedded within the taught curriculum through collaborative working practices between students, EAP and disciplinary teachers. I then go on to consider what this shift in perceptions and the positioning of language might mean for our understanding of what 'good teaching' (and, by definition, inclusive teaching) looks like. I argue that this change could help to place an institution in the 'competency internationalisation' suggested by Spencer-Oatey (2019) as the final stage of a higher education internationalisation process, where an institution has interculturally competent staff and students.

The previous chapters have presented an argument that language cuts across the planned, the incidental and the extra-curricula experiences of students in a number of ways which, I argue, need to be considered in curriculum development and planning. Currently, these areas are often disjointed, disconnected and viewed as the responsibility of disparate areas of an institution or of the student themselves without any external support.

Taking here the view that 'The curriculum breaks down the barriers between theory, research and practice ... the curriculum is a professional outlook, a practical philosophy of education' (Van Lier, 1996: 25), I maintain that it is through changes and shifts – through morphogenesis – within the curriculum, that institutions can move towards an inclusive and internationalised outlook. In this chapter, then, I draw together all the threads covered in previous sections of this book and begin to build a heuristic (see Figure 1, p. xi) around which to develop a curriculum that includes a focus on language as a central and important element for any discipline. I consider who is responsible for ensuring this focus on language takes place and what the implications might be for the pedagogical content knowledge that teachers would need to develop in order to build an understanding of the impact of language into their own disciplinary teaching practices.

Bridging the Content/Language Divide: A Heuristic

Throughout this book, I have considered different areas that impact on, and effectively become part of, the curriculum that need to involve an awareness of language and its impact. These are represented in Figure 1 (p. xi). I have already touched on the underpinning elements that act as a backdrop and on the six outer strands of this heuristic in the preceding chapters, exploring how each interacts with language, and how language use and misunderstanding can have an impact on the extent to which students feel able to access their HE curriculum. I argue that these elements should be conceived of as part of any HE teachers' Pedagogic Content Knowledge (Shulman, 1987) but that currently these aspects of a TPG student experience are largely disconnected. By drawing them together, reflecting on the impact and importance of each element, teachers can work out for themselves how to make changes in a way which leads to a fully cohesive, connected understanding of how language works as a rhizomatic thread through all teaching practices. Threading these through all learning experiences for all students we will enable the increasingly diverse student population to gain better access to the curriculum and all the different kinds of learning opportunities that are available in our HE institutions.

Connecting these six areas together more tightly into the curriculum, looking at where language differences can have both a positive and negative impact on *all* students' ability to learn, will be more likely to encourage students to access the facilities, support networks and learning opportunities they need. It may also encourage teachers and learners to connect and understand the 'potential of diverse or divergent pedagogical approaches and cultural and social capital within the global academic community' (HEA, 2014), moving away from the western or 'centre's viewpoint' (Holliday, 2004) of an internationalising curriculum. In other words, 'we need to ask whose version of sociology, engineering, medicine and so forth is being taught' (Pennycook, 1997: 263), yet at the same time 'accommodate the crucial reality that some knowledge is *in fact* better than others' (Moore, 2007: 34).

The stratification of knowledge is deeply implicit in our ideas of what education 'is' and what teachers 'are' and is also 'in principle a feature of any curriculum and any teaching' (Young, 1998: 18). Throughout this project it became clear that it was almost impossible for participants to stratify and separate language as a separate form of knowledge to other academic content learning (Paxton & Frith, 2014). It is also clear that language is not 'a neutral medium or conduit of information' but 'an inseparable element of disciplinary understanding and development' (Morgan, 2009: 88). Knowledge, practice, social context and the language used to define this are interconnected and while there is a place for a 'curriculum as fact' there also needs to be an understanding of the 'curriculum as practice' which is seen 'not just as something imposed on ... classroom practice, but as a historically specific social reality which teachers *action* and thus transform' (Young, 1998: 23). This is a curriculum viewed holistically, rather than from the point of view of a subject in isolation, with 'connective specialists' (Young, 1998) sharing a sense of the relationship between their specialisms. Despite moves towards interdisciplinary research and study, much HE teaching remains within 'subjects' and continues to be isolated, with modules maintaining separation between specialists. Language, however, 'plays a role in every discipline, not only in their textualisations but also in how they are taught and assessed. It is imbricated in epistemological shifts and theoretical frameworks. It plays a role as carrier of the past and mediator of future discourses' (Turner, 2011: 4). The impact of language as a real and integral factor in any disciplinary knowledge building is heightened when there are a large number of international/EAL students involved in learning, particularly if they are entering their studies with an IELTS or equivalent level of around 6.5. I suggest that it is language experts, or rather EAP practitioners, who could act as these 'connective specialists' building the sense of relationships between specialisms.

Thus, language consideration needs to become part of everyone's *pedagogy* and *content knowledge*; it should be highlighted where necessary *in* the content and seen as part *of* the content. Students need to be supported to move their learning away from 'language learning' or 'knowledge learning' towards a less defined separation, with language viewed as a means to communicate their disciplinary understanding. This language should no longer be seen as transparent, as a carrier of knowledge without holding any weight in and of itself, but as an integral part of the creation and communication of knowledge.

Language should become an explicit consideration in the varying *knowledge communication* demands placed on our students, and these demands should also be understood and not taken-for-granted elements of the learning process. Written and spoken communication need to be understood as social, dialogic practices that must be made explicit; and understood as part of the process rather than the culmination of learning (Lea, 2004; Ivanic, 1998). The culturally bound nature of the use of dialogue also needs to be unpacked. In classroom contexts, verbal participation should not be seen as the only evidence of learning and engagement; understanding that if a student is able 'to think about what they hear they will be as "involved" as when they answer questions' (Alexander in Leftstan & Snell, 2014: 73). Teaching can acknowledge the internal as well as the external dialogues that take place, but also needs to recognise when this is not taking place. In order to be able to identify the distinction between quiet but active participation and retreat into non-participation it is necessary to get to know students as individuals and develop an awareness around when and what barriers are being encountered, whether these are linguistic or otherwise.

Language should also be understood as holding potential to increase or decrease *cultural* and *social* capital and therefore be approached in ways that work to increase this or address issues around it for all – not just EAL – students. Language, and, in particular, academic language and ways of communicating are key barriers and deterrents for students from non-traditional UK backgrounds as well as for international students. Therefore, it is vital for Widening Participation and Access to HE initiatives as well as for the need for fully inclusive learning and teaching that language use is understood and considered by all as a key part of the curriculum.

Murray (2016a) has asserted that language is the common denominator in the four most valued of graduate skills – group or teamwork; leadership; intercultural and communication skills – again adding weight to the need to incorporate an understanding of how language can be manipulated to gain or lose capital. However, as Turner suggests, these communication skills are currently 'seen largely as a by-product of academic study ... rather than as something that may be addressed directly and intrinsically as part of the study process' (2011: 15). Interactions both inside and outside the classroom are negotiated to accelerate comprehension and production (Kumaravadivelu, 1994). Both students and teachers, English as a first, and English as an additional language speakers can and should be viewed as learners who challenge and collaborate with each other in 'pursuit of shared understandings of difficulties and shared ways of mastering them' (Land *et al.*, 2014: 214). In this way, interactions become intercultural in nature, and students are viewed as partners in a learning journey rather than as recipients of knowledge handed down.

Therefore, in my conceptualisation of a curriculum in which language is fully embedded, language is understood as troublesome *threshold* knowledge which requires journeys through liminal spaces in and of itself. In other words, language becomes part of every HE teacher's *pedagogical content knowledge*.

Collaboration and Co-construction

I have, so far, outlined the arguments as to why it is important to consider these different aspects of a TPG curriculum and how to thread an understanding and awareness of language throughout. In this section, I make some suggestions as to how institutions, staff and students can work towards achieving this language embedded approach to teaching and learning.

In focusing on the seven areas where language impacts on student access to and engagement with the curriculum, I do not aim to argue that the burden of teaching language within a discipline should now fall on the shoulders of the already over-stretched disciplinary content teacher. However, I do suggest that these teachers should take language into consideration when planning programmes, modules and individual learning sessions. As Haggis (2006: 530) has argued, 'the embedded, processual complexities of thinking, understanding and acting in specific disciplinary contexts need to be explored as an integral part of academic content teaching within the disciplines themselves'. The focus shifts from language *learning* to language *use* and from this consideration, content teachers should be then able to draw on the collaboration of those who do hold expertise in teaching language and understanding genre, discourse and the development of communication. Elbow (in Harwood & Hadley, 2004: 367–368) has argued: 'To write like a historian or biologist involves not just lingo but doing history or biology – which involves knowing history and biology in ways we [i.e. EAP teachers] do not'. This does not mean that there is not a place for EAP in this work, in fact the role of EAP becomes clearer. 'Given today's post-graduates and the communication challenges they face, ongoing EAP support for all post-graduate students is more important today than ever before' (Feake, 2016: 498). There should then at least be discussion between EAP and content teachers around how to ensure the burden of navigating the linguistic waters of a discourse is not left entirely to the students without even a life jacket to keep them afloat. This collaboration should begin with the confidence-building introduction to a content area, its language and discourse, provided by a language-heavy focused pre-sessional programme. It could then develop to involve the co-teaching and co-development of credit-bearing modules that connect language and disciplinary content more deeply and more explicitly, overtly acknowledging and rewarding the language learning that is currently a hidden part of, in particular but not exclusively, the EAL students' curriculum and learning journey as they undertake TPG study in the UK.

Collaboration between EAP practitioners and disciplinary academics may, at a basic level, simply involve a review and possible amendment to current learning and assessment materials or adding more specific guidelines or a glossary of terms to a module VLE (Virtual Learning Environment). The key is to ensure that language is highlighted explicitly, and instructions are clear and simple. This message is important not only when teaching international students, but also to ensure inclusive teaching practices for all, across the entire HE curriculum. It is necessary to build awareness of language as part of inclusive teaching practices in general and is therefore something to be highlighted in teaching development programmes and awards (see Chapter 9).

It is in the arena of *knowledge communication* that language proficiency or deficiency becomes most visible, and, for students, carries the highest stakes. It is a student's ability to communicate the knowledge they are developing that is ultimately assessed. It is therefore in this area that it is most important for collaboration between EAP practitioners and subject specialists to be developed, it is also here that there seems to be most reticence around developing any in-built teaching as extra to the current offer.

Some of this resistance is financial; as Wingate (2015) has suggested previously, investment in more teaching staff who are not viewed as core to disciplinary teaching can be a key barrier to full cross-institution collaborations. However, some concerns are also ethical, and knowledge based. For the majority of those involved in the project, the main concerns around knowledge communication lay in the written word. Thus, all student participants were largely concerned with the volume and complexity of the reading they were required to do, and with the pieces of written work that made up their assessments. In Case Study site 3, concerns were raised about students if they were unable to produce a written assessment that met expectations. EAP teachers generally expressed a desire to develop students' ability to produce written academic texts, although they were unsure whether the main purpose of their work was to consider the process of writing or the assessment of the product. EAP practitioners were also fearful of interfering with student writing for assessment in case this was viewed as providing an unfair advantage or encouraging forms of collusion. These concerns were similar to Harwood's findings from his research into the practice and knowledge of proofreaders, which revealed vastly different understandings of what was and was not acceptable (2018). Some of the teachers in the AHC site were less concerned with this aspect of knowledge communication; most argued that the students were clear about expectations in writing and were well supported through the process. For this site, poor linguistic expression, although making written work more time consuming to mark, was not seen of itself to prevent the communication of student knowledge.

International students likewise carry mixed messages around what kind of language knowledge is important, and often arrive in the United Kingdom with the belief that IELTS-based communication is adequate for academic study. This is demonstrated in some of the feedback received from students on the pre-sessional which suggested the programme was too focused on academic skills development and not general (IELTS) language:

- *In general, it [the pre-sessional programme] is different from the course I imagined and experience before. There are too many items about academic skills instead of language skills, for example, the part of Critical Thinking, Reference and reading circles. I am sorry to say that I do not feel my English improving through 10-week courses. … I think language course should concentrate on English skills just like two courses I took part in: Pronunciation and Grammar teaching. Maybe the system is a little far away from my dreaming class, I heard from my friends that language courses in University of *** and *** based on IELTS teaching. It is maybe more useful for students like us who are in lower English level, because our English is not acceptable and we are familiar with IELTS, we can understand and adapt to courses faster and better. (anon)*

Yet, later on in the data-collection period, Lin was able to bring together a range of concerns she had around her writing that highlighted the many areas of language, discourse, academic skills and content knowledge she felt she needed to develop and draw on:

- *I'm quite confident about the breadth of my reading, and I do have some of my own arguments in the essay. And I also have a clear structure. But I'm worried about the evidences to support my arguments and the depth of my understanding towards some literatures. And I'm also worried about grammar mistakes and whether the vocabulary I used is academic or not. (Lin)*

Several other students expressed regret that their language learning prior to arrival in the United Kingdom had been so focused on IELTS and grammar as they realised that it had not adequately prepared them for the demands of academic study and communication.

In this realisation, they were beginning to understand that a 'text is not a line of words releasing a single "theological meaning" … but a multidimensional space in which a variety of writings, none of them original, blend and clash' (Barthes, 1977 in Pennycook, 1996: 210).

These students, then, became aware that they needed to make a large leap from IELTS study to academic study and writing in English. I would argue that the curriculum needs to acknowledge this leap and work to address it, and that this can only be achieved through a partnership between students, EAP practitioners and disciplinary content teachers, and through a curriculum that highlights and acknowledges language as knowledge rather than language as a transparent and therefore ignorable carrier of knowledge. There remains, for example, a lack of knowledge as to what IELTS scores mean (Benzie, 2010; Murray, 2016a), and many academic teachers continue to work on the assumption that students who are admitted to their School do so with expert user (or 'native speaker' levels) of English language proficiency. When this is not the case and gets noticed, it is automatically viewed as the student's issue to sort out, either with or without a language teacher. Tripartite collaborations within the discipline in question would allow the EAP practitioner to bridge this lacuna of understanding and enable all to reach the understanding as outlined by Bruce (2008: 169) that:

> Achieving a discursive competence, in effect, is the ability to deconstruct, understand and reconstruct discourses in ways that are linguistically correct and socially appropriate, but also in ways which writers as individuals are able to achieve their own communicative purposes through their own authorial voices.

This focus on the discourse of a discipline, on how knowledge is created and communicated through language, rather than seeing language as a transparent conduit for disciplinary knowledge will require alterations or additions to the approach to teaching taken by both EAP and content teachers if they are to work collaboratively to bridge the existing epistemological gaps between the two areas. It will entail a shift away from educational specification and language 'knowledge blindness' (Maton, 2014). In many of the teacher–participant interviews within my study, the questions that participants raised highlighted a desire to maintain their traditional teaching practices, which mainly centred on seminar discussion, when the majority of the EAL (Chinese) student cohort did not actively participate. The emphasis here, then, is on the process of learning and knowledge *building* within this discipline, rather than the final product of knowledge communication. The view of language used here is not one of general proficiency or deficiency, but of its role in the cultural and social contexts of the University, where language and educational culture become conflated and teachers find that they are no longer able to continue a teaching practice that has developed via Lortie's 'apprenticeship of observation' (1975 in Tsui, 2003) where you teach as you were taught and learn as you learnt.

There is also space for much greater collaboration and co-construction of understanding with students through use of different languages. Whilst accepting that English is currently the common language of academic communication in most disciplines, and particularly in STEM subjects, and that my own study took place within the context of the United Kingdom – an English speaking country and institution –

an international University should work to place equal value on any language spoken. The case for English as a Lingua Franca in global academic conversations has been made with convincing force (Jenkins, 2013; Mauranen, 2003; Mauranen *et al.*, 2016), while Canagarajah (2011) has outlined the pedagogical possibilities of encouraging and supporting students to develop their translanguaging practices into written codemeshing, making use of their multilingual resources to more fully express their intellectual processes. While this may be a step too far for many contexts and disciplinary pedagogies, there does need as least to be a focused attempt to accept, enable and encourage students to feel comfortable when they make a practical choice to communicate understanding in a language other than English, and for its use to be seen as a draw on an extra resource rather than a shameful admittance of failure to meet communicative norms. The current position that all communication should take place in English; the insistence on elegant expression and the 'smooth read ideology which privileges the position of the reader' (Turner, 2018: 7) leads to a lack of clarity around academic practice. There is, therefore, little agreement and a range of (ethical) concerns around the extent of support for proofreading that should be available for students; this is now also being extended into questions over the use of translation tools. This is precisely because it is impossible to disconnect language from knowledge building and communication. There therefore needs to be a reflexive questioning of practice, of the pedagogy employed and the expectations placed on students, creating an 'ethical space... where the myriad issues of expectation, responsibility, time, and underlying rationales are openly debated, and nothing is taken for granted' (Turner, 2012: 24).

From this, it seems fair to conclude that 'Diversity in student population has necessitated some diversity in pedagogy' (Turner, 2011: 37). I now turn to consider what this diversity in pedagogy might look like, by considering content and language teaching within Shulman's framework of Pedagogical Content Knowledge (PCK), as this lies at the centre of my proposed heuristic for a language connected curriculum.

Developing Pedagogical Content Knowledge

The term pedagogical content knowledge (PCK) is used by Shulman (1987) to describe the interconnection between content knowledge and teaching 'know how'. This is the knowledge needed to understand how content is built and developed, where the cognitive load is greater, where learners are likely to meet sticking points and how to enable them to get beyond these. Shulman defines PCK as

> the blending of content and pedagogy into an understanding of how particular topics, problems, or issues are organized, represented, and

adapted to the diverse interests and abilities of learners, and presented for instruction. (Shulman, 1987: 8)

He separates the different elements involved in teaching into seven distinct but interlinked aspects: (i) content knowledge, (ii) general pedagogical knowledge, (iii) curriculum knowledge, (iv) Pedagogical Content Knowledge (PCK), (v) knowledge of learners and their characteristics, (vi) knowledge of educational contexts and (vii) knowledge of educational ends, purposes, and values (Shulman, 1987). PCK is, then, both the 'what' and the 'how' to teach combined to ensure effective student learning. Tsui (2003) adds to this understanding of PCK by incorporating Elbaz's (1983) five categories of knowledge: of the subject matter; of the curriculum; of instruction; of self and of the milieu of schooling.

PCK is generally used as a framework for explaining how teaching expertise can or has been developed, of highlighting how that intangible 'know how' is arrived at. There are clear links between Shulman's concept of pedagogical content knowledge and the notion of 'threshold concepts' (Land *et al.*, 2014; Meyer & Land, 2003, 2005) as a focus for transformative learning. Threshold Concepts requires 'expert practitioners looking back across thresholds they have personally long since crossed and attempting to understand … the difficulties faced from (untransformed) student perspectives' (Meyer & Land, 2006: 7); in doing so, there is a need to employ specific pedagogic skills. It is in the troublesome spaces of Threshold Concepts that the implicit knowledge of a disciplinary expert is made explicit by an expert teacher.

However, PCK has been criticised because 'it leaves unexplored matters of diversity and equity' (Dyches & Boyd, 2017: 478) as well as for failing to recognise that content and pedagogy are inherently bound, or, in other words, that concepts of knowledge are directly related to ideas about learning and teaching (Young, 1998). Moreover, the focus of PCK development continues to be closely connected to the development of disciplinary expertise, to ideas of knowledge being central in the curriculum, with little acknowledgement of who and how the knowledge agenda is set, and who is therefore privileged.

While taking the criticisms into account, PCK remains a useful heuristic for considering the different areas of knowledge and understanding required for good teaching practice. There are strong elements of PCK threaded through, for example, the core knowledge required for application to Advance HE's Fellowship Scheme. This scheme is interested in assessing the knowledge and values of those who teach in Higher Education in relation to the five core activities of learning design; teaching and learning delivery; assessment; creation of supportive learning environments and continuing professional development. Links can be seen here between PCK and the reflection-in-action highlighted in Schon's reflective practitioner model (1983), adding some of the precision and clarity that Eraut (2004) has criticised Schon's work for lacking. However, what is lacking in the scheme is any real guidance on *how* teachers develop this knowledge beyond engagement with literature and through experience. It works on the assumption that applicants are already involved in teaching practice and are able to evidence this within their application. They are required 'to order the mess' and 'verify their competence in an observable way [and these] observable, measurable professional skills may be valued over more ephemeral qualities such as empathy' (Webster-Wright, 2009: 718).

The professional body for EAP practitioners, BALEAP, has a similar scheme, based on the TEAP competency framework (2014) which claims to be a 'comprehensive statement of the knowledge and skills required by teachers of EAP' (Bruce, 2011: 104). However, the scheme has been criticised because 'moving beyond assimilating and reproducing to developing and transforming EAP praxis is absent from this framework, as is accommodation for, or recognition of, a more critically informed praxis and practitioner role' (Ding & Campion, 2016: 555). Moreover, there is no requirement in either award to focus on content specific language teaching. In the HEA award, it is likewise not explicitly required to provide evidence of teaching at TPG level in particular, or of an understanding of the demands placed on EAL/international students and how to address these via inclusive teaching practices.

While most HE institutions in the United Kingdom (and elsewhere) do offer support for teacher development, there remains resistance to engagement with this provision. Long-serving academics either do not prioritise this kind of training, or do not feel it is relevant to their work.

- *Training is difficult with academics because we're academics and we hate being trained. But the thing I think is really powerful is experiential stuff. If you could find a way to give us the experience of being an international student for a couple of hours, I think that would be really fascinating. (M10)*

For those who are early career/doctoral candidates and involved in teaching, whilst they do get support, they are also transient and working under precarious conditions. The institution may invest central resources to supporting their development as teachers but they do not always have easy access to those with disciplinary teaching expertise:

- *The problem arises because of constraints of time, resources, etc. A lot of the seminars are delivered by PGR, and they don't have necessarily the experience or training to do that. I think that's where learning and teaching can do a lot more. ... Because they do receive some training and they do receive – but it is not an easy fix because even if you fix it this year, next year you're going to have completely new people. (M3)*

Furthermore, formalised, institutional training and development opportunities, even when they do recognise that learning is socio-cultural, still compartmentalise and package professional learning into separate areas that can be generalised across the institution. The 'supercomplexity' of teaching in HE (Barnett, 2015) is broken down, and presented in the form of knowledge that is transferable across contexts, can be 'topped up', is cognitive, and can be acquired. This contrasts to the view that learning at and through work, which is more holistic and acknowledges complexity is more effective and is more greatly valued by teachers than attending a professional development workshop (Webster-Wright, 2009).

In order to develop effective pedagogical content knowledge in an increasingly complex HE landscape, it is necessary for teachers in a discipline to question their current practices but also to feel supported in the choices they make. I suggest that this can be best achieved through the development of microcultures (Roxå & Mårtensson, 2016), but that these microcultures must also involve a significant 'other' where the 'other' is outside the discipline and is therefore better able to question what are currently viewed as legitimate practices. Again, by working collaboratively together, an EAP teacher is able to become an outsider within, to disrupt and question current attitudes and norms, but also to become a trusted member of a network within a discipline that works to develop and transform student education and pedagogical practice. Concurrently, the EAP teacher should also become a legitimate member of the academy and, through interaction with content teachers, further develop their own authority, content knowledge and ability to de-code disciplinary discourses.

Pedagogical content knowledge for content teaching

The need to engage a broader range of students, from a variety of linguistic and educational backgrounds requires changes to practice and a rethinking of approach; the pedagogy of osmosis (Turner, 2011) is not enough. Teacher–participants in my study felt that their traditional pedagogical content knowledge for teaching at TPG level was being undermined. This was most clear in the AHC site, where there was a rapidly changing and diverse cohort of students. Teachers felt that they needed to consider more than the subject matter and their own disciplinary identity. Their traditional pedagogy followed the Socratic/apprenticeship model where knowledge is developed via discussion with an expert knower. Teaching made the assumption that students at this level are successful, have already developed the required habitus to work within their field and do not therefore need to be taught how to learn, but would simply be able to participate in intellectual discussion and debate. This was being challenged by the students they were now encountering.

Thus, teachers reported needing to provide guided questions for seminars to encourage students to pick out key points. Bringing in elements of competition was another tactic used to encourage participation, as was use of the VLE to provide extra guidance outside seminars and lectures. As experienced teachers, then, there was evidence of reflection around 'an integration of knowledge of subjects with knowledge of students, contexts, curriculum and pedagogy' (Tsui, 2003: 58). The perceived difficulty was in doing this in a manner that maintained the value and rigour of a TPG award while ensuring inclusion of all.

Content teachers felt isolated in their teaching practices, and were unsure as to whether their own, and the practice of others was effective or not. Their pedagogy was based on their own experiences and preferences; they lacked access to the significant network that Roxå and Mårtensson suggest enables teachers to develop their practice:

- *Now I don't know enough about that because I, one of the awful things about universities is we never see each other teach so I kind of guess, you know? Some of my colleagues I think god, you know, I can't imagine being taught by them whereas some of them, I think must be quite good. I only know what I am like. And I try to teach as if, in a way that I would like to be taught myself. (M7)*

Good teachers, from both their own and their students' comments, were perceived to be those who demonstrated an ability to ground theory in everyday practice, transforming their knowledge of subject through explanation or representation into a comprehensible form that related to their (at least assumed) knowledge of their students' interests.

- *when I then connect with the lecture I'm really inspired by the teacher because the teacher will maybe connect a point with other different evidence and the different other points and sometimes they will find some really critical question to ask us to answer it but we think it's really good and inspires us to find some connection between different things. (8L)*

The ability to do this was also connected to classroom organisation as well as clarity of explanation. This was not simply choice of seminar style layout or use of laboratory for practical demonstration and experimentation, but also connected to seating plans and knowledge of students, linking content and pedagogical knowledge together. However, what these teachers were not doing was developing an understanding of language as knowledge and knowledge creation. There remained a tendency, as reported by Chanock (2007: 273) for academic teachers to separate disciplinary knowledge (and therefore teaching) from academic literacies (teaching). In fact, as these teachers explained, they felt that currently they needed *to resist* a shift into language teaching.

- *I think it's neither practical nor appropriate at the Masters level for academic assessors of the work to be doing that [commenting on language] ... I think it's their responsibility to seek out that expert help'. (M5)*
- *I'd be thinking all the time that I cannot spend time telling them how to write all these sentences better. I can't and I feel bad about it. I'd be saying about language and phrasing, but I wouldn't really explain more than that about what I mean, except maybe once or twice. I'd then say that this is something you could get some help with and then point them to the languages support. (M1)*
- *I'm not an expert on language, I don't know if I'd call it impeding language, but things that made me not be able to follow what they were saying, that comes from a misunderstanding of what they were reading. I don't know if that's the same thing. (M1)*
- *we probably don't give enough feedback to say, you know, well how can that, but then I think if it gets to that level it's beyond our expertise. So we have to draw the line somewhere. We're not language teachers. We will struggle to articulate to the student, we certainly don't have the time to articulate to the student. (M4)*

This resistance and denial of expertise needs to be addressed if, as outlined above, the use of language (in this case English) is seen as key to enabling access to disciplinary knowledge, and is also, therefore, one of the barriers for the growing number of students who are studying with English as an additional language. I suggest that one way of addressing this is for language teaching pedagogies to be adopted as part of a pedagogical repertoire for those teaching in other disciplines.

There are techniques and methods that can be used to encourage greater student participation in classes. For EAP teachers, these are a common, taken for granted, element of their pedagogy. As such, it has become tacit knowledge and hard to describe:

- *Well, I think because we are used to working with international students, we might have a manner and related behaviours which are more conducive to them learning somehow. All these extra little bits and pieces. (LC1)*

These 'extra bits and pieces' are the result of training, qualifications and experience in (English) language education, which focuses on encouraging communication, developing the ability to listen and participate, often in an intercultural environment. A number of student–participants who had studied in both the EAP unit and then a different

School expressed confusion over the non-transfer of these communicative skills from their EAP pre-sessional as they moved onto their TPG programme. Observations of classes showed this transfer most successfully occurred when content teachers employed pedagogical techniques that, while remaining true to those of their own discipline, also incorporated those of the language classroom.

By working together to share pedagogical practices and considering language development as an integrated element of an inclusive curriculum, it may, therefore, be possible to address some of the issues around non-participation of (EAL) students.

It could be argued, though, that following my suggestion of embedded content/EAP collaboration, it is the PCK of EAP teachers rather than content teachers that needs to focus on an understanding of language as knowledge. In order to support students within a specific discipline, it is important that EAP teachers are at least aware of signature pedagogies of that discipline. However, it is even more vital that EAP develops its own signature pedagogies, based round a clear but extensive knowledge base, that moves beyond those the majority of EAP teachers are trained in via their initial language teacher training courses.

Pedagogical content knowledge for EAP teaching

Just as content teachers were feeling challenged by the student reaction to (or non-engagement with) the approaches to teaching – the pedagogy – they employed, EAP teachers faced similar challenges as their teaching practices moved further away from isolated language and skills teaching towards a more content-driven approach.

In English as a second language (ESOL) teaching, pedagogical content knowledge has been defined as the ability to connect knowledge about the language system (including phonology, lexis, grammar and discourse) through teaching of the four communicative skills and language learning strategies (Tsui, 2003). It is this ESL view of language learning theory, where language develops via a process of acquisition of grammar, lexis and phonological forms, that is the starting point for the career of most EAP tutors. Gee argues that 'when English language learning is regarded as simply developing skills it becomes an individual activity and responsibility rather than a complex process of interactivity in various context' (2008: 31). What is needed then, is a move towards the ability to apply and use this language for different purposes, where language becomes inextricable from context; in other words, a socio-cultural view of language use (cf. Kramsch, 1993, 1998, 2009; Norton, 2000; Norton Peirce, 1995).

I have already discussed the confusion over the required knowledge base of those involved in teaching EAP. Some EAP teacher–participants remained stuck and unhappy, feeling that they were not giving the students what they wanted, and not addressing basic language learning needs. Others, even over the short period of a six-week pre-sessional programme, demonstrated a developing understanding of where their linguistic expertise should and could connect with the disciplinary content knowledge needs of their students. Interaction with disciplinary texts and genres and attendance at lectures enabled them to engage in participatory social enquiry and see the knowing and the language in practice. This, in turn, allowed them to consider where the main conceptual and linguistic barriers may be for their students. However, from this understanding some then found that they lacked a sound knowledge base within their own discipline, EAP, on which to draw. Although they were able to see what needed to be done, they were not fully confident in supporting their students in this new way. This resulted in frustration, a decreasing sense of authority and retreat into teaching either discrete skills or too much focus on the content itself.

Content teachers could learn much from EAP in terms of pedagogy, intercultural communication and an approach to education that privileges student voice, emotional well-being and confidence building. Conversely, EAP teachers can learn much from a collaborative relationship in terms of an approach to knowledge. This includes an understanding of the disciplinary epistemologies that an EAP practitioner needs awareness of in order to support their students. However, and more importantly, by being able to fully locate language within the reality of a TPG curriculum, it allows EAP teachers to identify the gaps in their own knowledge and encourages them to investigate the best ways in which to address these so that they can meet the real, disciplinary literacy needs of their students rather than the theoretical, disconnected needs they have previously built their teaching around. Within this relationship there lie real possibilities for meaningful contextual investigation and exploration of the different knowledge bases that EAP is able to draw on. The EAP teacher, whose identity is fully embedded in student education, benefits from the opportunity to develop a SoTL project of real benefit to both themselves and the institution, adding and building strength to the position of both. The content teacher, who may not have time for SoTL when understood as pedagogical research, can benefit from the same SoTL work as a form of academic development (Geertsema, 2016).

Chapter Summary and Recommendations

Despite difficulties in separating language as an expression of content knowledge, the data suggests that content teachers were not confident or had not previously considered how to make the language in their content teaching explicit, supporting Wingate and Tribble's claim that 'there is evidence that academics are unwilling to get involved with … or lack the confidence to offer this kind of support' (2012: 488). Conversely, EAP teachers lacked confidence in drawing out and developing language through the complex content of the students' discipline. Before considering a language-embedded curriculum as suggested above, there therefore need to be development opportunities for staff to enable them to approach this area with relative confidence and competence.

Again, it seems that there is a strong argument for collaboration between educational development, EAP and content teachers to encourage a sharing of knowledge and good practice. Learning through experience that is directly applicable to an individual teaching situation appears to be the preferred method of teaching development. However, experiential learning does need to be re-enforced with theoretical knowledge and evidence of its value. Developing observation opportunities across the institution as well as within a School would allow further reflection on the different pedagogies in place and an opening up of conversations around what inclusive teaching might and should look like in different areas of the curriculum, as would an open, cross-institutional mentoring scheme with a focus on student education.

This sharing of practice needs to be an on-going process. It needs to include those who are early on in their academic careers as they frequently take on teaching responsibilities alongside their research but should also encourage more established teachers to reassess and review their current practices. Ultimately, the development of inclusive, language aware teaching practices needs to be an accepted norm within the training and development of good HE teachers across all disciplines, with support materials and opportunities for continuous professional learning (CPL) readily available.

8 Implications

In this final chapter, I suggest that the responsibility for ensuring students are able to access the curriculum in Higher Education, at all levels of study and regardless of the amount of time they will spend studying, lies firmly with the Institution. This is true for students from all language and cultural backgrounds, and with any range of other needs. Students should not feel that they are in deficit because of individual differences or, as when considering expert knowledge of other languages, their collective differences to the privileged norm.

Perhaps the strongest theme that emerged from my study is one of individual academic teachers and EAP teachers working hard and to the best of their abilities to find a way to support students through their programme of study. However, they are doing this in degrees of isolation, without the support or training that would afford them the time or space to develop a clear understanding of how to work with students in an inclusive and content/language aware manner. Concurrently, they are working within a greedy system where the number of students, and therefore the number of students with diverse needs, is increasing exponentially. Staff feel even more time poor and even less well equipped to provide the individualised support they would like to give to their students. Students who are relatively transient, only within the system for (less than) a year, struggling to adjust to new systems and languages are generally silenced and voiceless.

As Turner has argued, work around language and 'languaging' has generally been 'underestimated, undervalued and marginalised in the institutional discourse of Higher Education' (2011: 4), and is only noticed when the message a student is trying to convey is not successful. At this point, language *is* noticed and *is* viewed as needing remedial attention. In this way, the student is positioned as being in deficit, and the EAP unit to which the student is sent is positioned as providing a remedial, lower-status service. However, because of the diversification of student population and the work involved in educating them, language, its power and its contribution to disciplinary discourses has now become an issue worthy of consideration within Higher Education. More broadly, social and political public discourses have also highlighted inequalities that grow from powerful English-language speaking hegemonies. Within HE itself there is an increasing number of Universities in non-English speaking countries that are developing programmes to be delivered through English as a Medium of Instruction (EMI) as well as a continuing interest in the development of trans-national campuses that promise to deliver, for example, a University of Liverpool experience in China (Xi'an Jiaotong-Liverpool). All of this suggests there is now urgent need for HE institutions to develop an institution wide understanding of the role played by language in all their activities, and to develop a well thought out and nuanced strategy and policy that focuses on language in order to highlight where support, development and access needs can be provided and met for both students and staff. While the 'empirical reality of language and intercultural communication lends a relentless practicality to the issues that need addressing' and means that those involved in the struggle around language and discourse are by necessity already finding ways to navigate this reality, it is now imperative that languages and language issues are foregrounded. In doing so, language then 'becomes a matter of institutional politics' (Turner, 2011: 2–3).

Here, then, I consider what the implications of this foregrounding of language in the HE TPG curriculum means for institutional policy and strategy.

Policy and Strategy

The suggestion for collaboration between EAP practitioners and disciplinary content specialists is not new. Several researchers and practitioners in EAP and Academic Literacies have already suggested that this model is the ideal. Murray (2016b) and Turner (2011) have also argued forcefully, but largely theoretically, for a shift in institutional thinking about language and a move towards foregrounding it and the EAP practitioner across the curriculum. Wingate and Tribble (2012) outline how a genre-based approach to academic writing could be enacted to support students within their own disciplines, but do not suggest how to establish the working collaboration that would enable this to happen. Wingate (2015, 2018) has argued further for this kind of collaboration and provides an example of successful practice working with students on an MA Applied Linguistics. Her conclusion, however, is that rolling this out across a University would be too financially restrictive. In terms of available literature, the most adoptable and adaptable example of this kind of collaboration-in-practice is provided by Alexander *et al.* (2017), who outline what they term the CEM model (Contextualised, Embedded and Mapped) and how they implemented this within a business programme. To date, however, the literature suggests that moves towards this collaborative relationship are based on individual working relationships and interests, rather than a fully embedded University policy and approach.

Here it is important to acknowledge that available literature does not provide the full picture. The divide between research and practice in EAP is well documented (Ding & Bruce, 2017; Hamp-Lyons & Hyland, 2005; Hamp-Lyons, 2011), and collaborative teaching inside an institution seems to lie within the realm of practice rather than research. A search of University websites, occasional presentations at BALEAP conferences and mailing list conversations suggest that there are already extensive and established collaborations between EAP practitioners and disciplinary academics; it is heartening to see the recent publication via BALEAP 'Insessional English for academic Purposes' edited by Brewer *et al.* (2019). These collaborations are particularly evident in institutions that specialise in specific disciplinary areas (for example Imperial College, London or Goldsmiths College, London). In the United States of America, Hartig's (2017) book suggests that EAP practitioners (or the US equivalent) have gained some access to working within a disciplinary area, as does Kettle's (2017) work – although this again seems to focus on one particularly invested academic colleague. There are also examples of collaborative work between Academic Literacies and content teachers in a range of contexts (see Lillis *et al.*, 2015), but again with little evidence of strategic, institution-level implementation.

The Department of Education, Employment and Workplace Relations (DEEWR) in Australia produced a useful guide (2009) that suggests good practice principles for English language proficiency for international students and in the UK, the British Council (2014), the APPG Inquiry Report (2018) and the #WeAreInternational community all push for international inclusion and integration. It is important to note again here, though, that the label 'international' and English language proficiency should not be conflated. Nor should policy, or policy suggestions, be accepted as enough. It is necessary to ensure that this policy can be translated into practice, and practice that is effective. The documents created by these organisations can be used by leaders of EAP centres to persuade other university leaders of the importance of this work, while offering collaboration as a practical step to delivery of such a policy. For language policy, and the central question of a student's ability to communicate with their academic peers, it is necessary to have a policy that moves the learning of EAP away from 'ivory ghetto of remediation' (Swales, 1990: 6) to something that is available and useful for all and that goes beyond developing proficiency in a language towards enabling access to knowledge and developing expertise in academic communication.

Such an institutional policy would need to consider the interconnectedness of language, and how it threads through all aspects of a student's life and the University curriculum. It has an impact on conceptions and understanding of Academic Integrity (see Bretag, 2017) and therefore on approaches to editing and proofreading (Harwood,

2018). To date the policy focus has been on improving English language proficiency in order to assimilate EAL students into the language of privilege. I suggest that consideration needs to be given to how this one particular language is privileged over another and to what extent the use of other languages, can (and should) not only be tolerated but valued and encouraged, allowing students to draw on all resources available to them. This in turn impacts on policies in relation to the acceptance, or not, of digital tools that provide language support or help, for example translation software. The social, cultural, emotional and embodied aspects of language use and development need to be considered in such a policy and connected into the different support systems available to all students as well as the academic context in which they are working.

Spolsky (2012) describes the three components of a language policy as including the actual language practices that occur within a community, the values placed on the different forms of language in use and the management of language use – where those with perceived power attempt to manage and control the boundaries of acceptable language use. The current value placed on linguistic accuracy and the sophisticated use of English expression is problematic; there has been little change since Cameron wrote in 1995 of 'linguistic bigotry' being the last remaining publicly acceptable prejudice. Murray's account of academics bemoaning the deterioration of written work due to poor language proficiency (2016a) is echoed in many institutions and in the press. While many of the participants in my study argued that they were able to see beyond poor language, language was still being measured as poor, and as finite, rather than as an ongoing process. In this way, one person's language continues to be seen in deficit to an idealised norm.

In order to create an institution that is inclusive and works towards an international campus rather than a UK-centric campus that includes international students, it is necessary to create a policy that values all forms of language in use and is able to see how these different forms can be used to express complex and intelligent thought processes. The management of such a policy should include the development of a clear understanding of what this would mean in practice, of how it would be enacted through the curriculum. It should also involve an investment in the staff who would be required to change their approaches to language and knowledge communication.

Approaches and Practices for Teaching and Learning Within an English-Speaking Environment

An institutional move towards a more international and a more inclusive approach to learning and teaching involves much more, therefore, than writing a policy. Within most UK universities, and most Faculties and Schools, there will be a required shift in culture and a questioning and rethinking of practice across the curriculum. This in turn will require reflexivity on behalf of all those involved, coming through reaction and interaction with students and staff across the institution. Because 'the values of the teachers mediate the policies of the institution' (Canagarajah, 2011: 416), and because the policies of the institution involve great levels of complexity and uncertainty, it will not be possible to implement any policy without the understanding and continuous learning of the teachers involved.

It is possible to show visible investment in supporting staff who are engaged and working towards a more inclusive and internationalised approach to their teaching through awards events that celebrate individual impact in these areas. This highlighting of good practice does have some impact on changing culture. However, it is investment of time, over long periods, with a shift towards continuous professional learning (CPL) that brings around culture change. Ding's (2019: 70) argument in relation to EAP practitioners can be extended to all HE teachers:

> it is imperative to invest in practitioners so that they can achieve an identity, through development and scholarship, that affords them the recognition and agency needed to fully participate in and influence the university.

Research suggests that the most effective CPL takes place through contextualised experiences that provide opportunities for reflective action (Webster-Wright, 2009). For this to be authentic and meaningful, it needs, therefore, to take place within the workplace, in action rather than through pre-planned and standardised development sessions. The institutional investment, then, comes from enabling staff to engage in conversations which encourage learning to take place. Berg and Seeber (2016) argue for a greater value to be placed on time spent in conversation, discussing and considering different perspectives. Their suggestion is that much quality has been lost through the increased intensity of work in Higher Education and that this can, to an extent, be regained via 'slow' conversation. Webster-Wright also connects the development of communities of practice (Lave & Wenger, 1991) to the development of a stronger identity and the ability to proactively act to alter difficult situations. Roxå and Mårtensson (2016) suggest that significant professional learning about teaching takes place via microcultures that develop between small groups of trusted others who can provide support and enhance understanding.

This approach to professional learning is echoed within language teaching research. The use of Exploratory Practice, where teachers collaborate to explore puzzles within their own practice, has been suggested as a means of professional learning that is inclusive of all those involved (students and teachers) and works towards enhancing quality of life (Hanks, 2017). Within EAP specifically, 'opportunities for longer-term, on-going development initiatives are what teachers find to be most valuable. Amongst these types of development initiatives, a recurring point from the research is teacher participants' comments about the value of informal learning opportunities' (Ding & Campion, 2016: 554).

Given the respective strengths and difficulties that the content and EAP teacher participants reported in terms of their professional practice, it seems clear that an institutional strategy that encourages and enables much closer collaboration between the two would work not only for the direct benefit of students, but also provide opportunities for developing 'significant networks' for professional learning amongst staff.

For EAP teachers, the benefits of this kind of collaboration are clear. EAP acts, generally, as a bridge between students and lecturers, between language and content knowledge. However, the student destination is not always clear to the EAP teacher without the collaboration of content teachers.

EAP is often accused of being accommodationist, of applying and enforcing the rules of others rather than creating their own. These criticisms, coming from Critical EAP (Benesch, 2001) and from Academic English as a Lingua Franca (ELFA) (Jenkins, 2013), suggest that EAP needs to work to transform rather than socialise students into the academy. However, in order to do this, EAP teachers need to fully understand what the current norms, the epistemologies, the cultures and values of different disciplines are. Having clearer knowledge of the localised disciplinary norms from within would allow the EAP teacher to work with their students with greater knowledge as to what it is reasonable to question and to change. In this way, teachers can engage in Critical EAP practice that goes beyond that suggested by Benesch, where the EAP practitioner still maintains control, and therefore power, over the subject of study and begin to work with students as they explore language within a subject of their own choosing. By working within a discipline, the EAP teacher becomes a recognised member of the academy, with equal status to their content-specialist counterparts. They acquire greater access to the content knowledge (through access to lectures, seminars, Virtual Learning Environments as well as staff meetings and corridor conversations) that is held and valued by the discipline, not just the written or spoken product – moving beyond the paradigm shift of the ELFA model proposed by Jenkins (2013), towards a practice that interweaves language, literacy, discourse practices within and through the development of disciplinary knowledge.

However, in order to engage in this kind of work, it is necessary for EAP to examine itself carefully and begin to make a shift away from the focus on teaching-only credentials. While the DELTA may train teachers, it does not prepare them for the kinds of academically rigorous, discipline-specific needs-based work I suggest here for EAP. Given the current contexts under which EAP units operate, it is difficult to identify which move should come first. It is necessary to address the structural issues that surround EAP practice, to consider who teaches EAP, their trajectory into the profession – both in terms of prior experience and qualifications – where units are located and the continuous round of teaching which perpetuates inequality with other academics in terms of workload balance away from scholarship or research. This shift needs to be pushed both from within the EAP community but also written into institutional plans and policies. It is not easy to do this when faced with the economic pull of 2000 international student fees on a summer pre-sessional and the need to staff enough classrooms to accommodate them.

However, given the right conditions, strategically planned collaboration across the university could enable EAP practitioners to work towards developing the domain of knowledge that Ding and Bruce suggest is currently most lacking in our practice – 'a sociological understanding, not only of the disciplines and departments, but of the university' (2017: 207). Through scholarly enquiry that is contextually situated and relevant, EAP teachers may find themselves enabled to question, critique and hopefully transform practices in collaboration with their disciplinary colleagues and their students, and thus feed back into addressing the structural issues that currently constrain EAP.

Conversely, through discussion, collaboration and sharing of practice, the EAP teacher can help the content teacher develop a clearer understanding of how to encourage direct participation and communication, how to make assessments linguistically clear. Feedback on, or co-delivery of lectures could, as Jenkins (2013) suggests, highlight use of metaphor, for example as a barrier to understanding. This work would clarify areas of linguistic difficulty that are the responsibility of teachers to consider, not of students to understand, developing a clearer understanding that some language is more culturally bound than others.

Together, through shared experiences and collective action, I suggest that EAP and content teachers can work to transform and develop their practice so that it integrates a focus on language within the development of disciplinary knowledge and understanding. In turn, TPG education practice could then become more inclusive of all. It would (admittedly rather idealistically) shift the development of a curriculum towards becoming a space

> where teachers and students can go *between* and *beyond* socially constructed language and educational systems, structures and practices to engage diverse multiple meaning-making systems and subjectivities, to generate new configurations of language and education practices, and to challenge and transform old understandings and structures. In so doing, orders of discourse shift and the voices of Others come to the forefront. (Li, 2018: 24)

Conclusions

[A]ll instructional manoeuvres are politically charged and therefore never neutral (Dyches & Boyd, 2017: 476).

Teachers mediate between institutional politics and reveal their own commitments in the way they interpret and enact them. Making the choice to consider or ignore language, which is perhaps the most evident discursive practice in which privilege and prejudice are woven into our everyday lives, is therefore a political decision. In arguing for language to be considered across the curriculum, I am arguing for an inclusive approach to learning and teaching that

> necessitates a shift away from supporting specific student groups through a discrete set of policies or time-bound interventions, towards equity considerations being embedded within all functions of the institution and treated as an ongoing process of quality enhancement. Making a shift of such magnitude requires cultural and systemic change at both the policy and practice levels. (May & Bridger, 2010: 6)

Thus, while acknowledging the struggles that many students face with language (and this includes those with, for example, dyslexia as well as EAL learners) it is necessary that they are not treated inequitably as a result. Studying already takes up more of their time than for their peers; it is hardly surprising that they cannot find the time to attend, or that they often fail to make the connection between, adjunct sessions that aim to provide support but have little connection to the students' core learning priorities. It is this method of provision which should be seen as being in deficit, rather than the students themselves.

It is here that the importance of agency is highlighted. While it is the structure and culture of HE, the provision and the curriculum, that constrains and excludes, that structure and culture is and can be elaborated by the agents who interact with it. Much research

> has focussed on how international students can be 'enabled' to succeed academically rather than how they might influence curricular enhancements ... the resultant epistemology has both contributed to the application of the deficit model when teaching international students, and silenced home students' voices, ignoring their diverse academic and personal needs. (Ippolito, 2007: 750)

International students and policies of internationalisation are already changing the global Higher Education landscape. If we are who we are because of 'what we care about most and the commitments we make accordingly' (Archer, 2003: 120), we need to focus our attention on care for our students and to give them a voice. By acting for them and with them, the structure and culture can shift towards one of inclusion and consequently of social justice when social justice is viewed as 'as an active commitment to dismantling … hegemonies and the structures that both represent and perpetuate them' (Bell, 1997 in Dyches & Boyd, 2017: 478).

If the language requirement is to remain the same for university entry, the curriculum (as both a structure and an agential actor) needs to work to meet the language development needs with the academic content learning in a much more connected and holistic manner. International students, just like all other students, should be able to gain full access to the curriculum. To do so, their needs need to be taken into account but viewed within a higher education 'premised upon the explicit aims of inclusion and diversity' (Lillis, 2003: 192). It is necessary, then, to make a clear separation between the understandable struggles with language that students who enter at the minimum language proficiency level are going to face, their academic abilities and their emerging understandings that are frequently wrongly blamed on but enmeshed within the language they are required to use. International students with an IELTS or equivalent score of 6.5 are, *in fact*, in deficit linguistically in terms of the language that will be their medium of instruction. Although achieving the required IELTS grade, they do not in reality necessarily meet, to use the language of current governmental education policy, the 'standard expected' language level for TPG study. They are unable to express and communicate their thoughts as eloquently and accurately as they would wish to in English. To pretend otherwise is to reject their own feelings as being invalid and deny the need to provide the necessary guidance. Students *do* want to understand the norms of a discipline and know the language used within it; only those who are aware of those norms and have been through some element of being socialised into a community will develop the capital that allows them to fully express their voice and potentially change the discourse in the manner suggested by the Academic Literacies or ELFA movements. In fact, part 'of social justice knowledge is understanding everyday discourse … as the mechanism through which biases are perpetuated in both the general structure of language and the ways it is used to construct myths and justify norms' (Dyches & Boyd, 2017: 481). Through this understanding, students can begin to take an active role in their own education and teachers can begin to deconstruct and alter the norms so that all students have the ability to interact on an equal footing with their peers and achieve 'participatory parity' (Leibowitz & Bozalek, 2016).

Afterword: The Engaged Scholar

There is a necessary tension throughout this book between the specific context and the need for others to be able to generalise from this. I would argue that storytelling, the narrative of a single story, if told with empathy, depth and an understanding of the multiple lenses it is possible to view the story through, enables a story to resonate beyond the individual or the local. So here, I return to my own story in the hope that others can take something of value from it.

In Chapter 1, I described my own journey through scholarship to the point of writing this book. In doing so, I described the journey as 'accidental'. I applied for a role I had not really expected to get and consequently found myself in a very visible position as I undertook my first major scholarship project, with no real training or formalised support.

While I was aware of the luxury of time afforded by my shift into a centre for teaching excellence, and felt myself to be in a privileged and lucky position, I also felt fearful and, at times, resentful of the external pressures being put on me to go public before I was ready to do so. I also felt guilt for feeling fearful and resentful when I knew how lucky I was.

I also felt, and continue to feel, the weight of representation. As an EAP practitioner who had almost overnight moved out of the margins into an institutional spotlight, I felt the need to demonstrate that as a profession we are worthy of playing a central role in the academy; that the teaching and scholarship we engage in has value across campus. However, what I was seeing in my data was that many of us do not feel ready to play that role. I included myself in this conclusion – it was over 15 years since I had completed my Masters degree and since then my scholarship work had been small scale and very practice based and practical – I was, in fact, an average EAP teacher, with some management responsibilities and in the lucky position of having a full-time permanent post. Yet there I was, being required to attend meetings across campus and speak with institutional authority about language, content and the international student.

What became clear though was that although I felt I didn't know enough, I did already have some knowledge, and definitely had ideas, opinions and experience that others with more institutional power did not, simply because I was approaching an issue from a different perspective. Through being forced to voice my thoughts publicly, they developed quickly. The desire to back up my ideas with greater authority (i.e. to not look stupid!), as well as the external expectations to produce something, pushed me to continue with my project and built my confidence. It also kept me focused and has continued to do so for three years now, as the ripples of my project recommendations grow across my institution. It is this continued and sustained focus on scholarship as impactful, carrying authority and feeding back into practice (whether that be individual, classroom, or policy) that, for me, should define what an EAP Practitioner ought to be: someone who views teaching, learning and scholarship as inseparable elements of their professional identity. In a context where precarious contracts proliferate and EAP remains at the margins, it is through this practice, supported by evidence and our own knowledge bases that allow us to speak with authority and move towards playing a more central role in Higher Education. I am still fearful and will probably always suffer from imposter syndrome. However, my scholarship is no longer accidental; it is engaged and purposeful and I am beginning to see the impact it has on those I care about most – my colleagues, our students and the wider EAP community.

Notes

(1) This label has also been criticised for denoting deficit. However, in this context, it does at least suggest, through the term additional, that students are adding, and therefore offering extra, rather than the reductionist nature of 'non' in 'non-native'.

(2) ELTS (International Language Testing System) is the most commonly recognised international language test in UK HE. Whilst there are a range of other recognised tests (Pearson; Password; currently not TOEFL in the United Kingdom), most of these are always measured in equivalency terms to IELTS scores as the most commonly understood.

(3) The Common European Framework is a scale of proficiency measures in all four language skills that language learners of any language are encouraged to measure themselves against and use as a form of self-assessment. The scale is also now often used as a benchmark for assessment in more localised language assessments.

(4) Thank you to Alex Ding for this insight.

(5) The unit also provides some credit bearing modules both within the School and for the International Foundation Year. Some of these students are 'home' students who are proficient only in English. Other teachers work on General English programmes.

(6) I do acknowledge that well-known researchers can draw students to choose a programme, and that the title of professor does hold a great deal of cultural capital that can lead students to presuming an excellence in teaching as well as research.

(7) Note I do not use 'language' teacher here, as EAP should, in theory, include teaching more than language. See discussion in Chapter 7 around what it means to work in EAP rather than ELT.

References

Adams, R. (2017) See https://www.theguardian.com/education/2017/jan/12/tougher-stance--overseas-students-could-cost-uk-2bn-a-year-higeher-fees-brexits-student-visa-numbers The Guardian Online (accessed January 2017).

Adichi, C.N. (2009) *The danger of a single story*. TED Talks posted 7/10/2009. See https://www.youtube.com/watch?v=D9Ihs241zeg.

Alexander, O., Sloan, D., Hughes, K. and Ashby, S. (2017) Engaging with quality via the CEM model: Enhancing the content and performance management of postgraduate in-sessional academic skills provision. *Journal of English for Academic Purposes* 27, 56–70.

Allwright, D. and Hanks, J. (2009) *The Developing Language Learner: An Introduction to Exploratory Practice*. London: Palgrave Macmillan.

Amadasi, S. and Holliday, A. (2017) Block and thread intercultural narrative and positioning: conversations with newly arrived postgraduate students. *Language and Intercultural Communication* 17 (3), 254–269.

APPG (All-Party Parliamentary Group for International Students) (2018) Inquiry Report A Sustainable Future for International Students in the UK. See http://www.exeduk.com/wp-content/uploads/2018/11/APPG-Report-FINAL-WEB-1.pdf.

Archer, M.S. (1995) *Realist Social Theory: The Morphogenetic Approach*. Cambridge: Cambridge University Press.

Archer, M.S. (1998) Realism and morphogenesis. In M. Archer, R. Bhaskar, A. Collier, T. Lawson and A. Norrie (eds) *Critical Realism: Essential Readings* (pp. 356–381). Abingdon: Routledge.

Archer, M.S. (2000) *Being Human: The Problem of Agency*. Cambridge: Cambridge University Press.

Archer, M. (2003) *Structure, Agency and the Internal Conversation*. Cambridge: Cambridge University Press.

Archer, M., Bhaskar, R., Collier, A., Lawson, T. and Norrie, A. (1998) General introduction. In M. Archer, R. Bhaskar, A. Collier, T. Lawson and A. Norrie (eds) *Critical Realism: Essential Readings* (pp. ix–xxiv). Abingdon: Routledge.

Baker, W. (2016) English as an academic lingua franca and intercultural awareness: Student mobility in the transcultural university. *Language and Intercultural Communication* 16 (3), 437–451.

BALEAP (2014) TEAP: Accreditation scheme handbook. See https://www.baleap.org/wp-content/uploads/2016/04/Teap-Scheme-Handbook-2014.pdf.

BALEAP JISCmail (2017) *Academic' and 'Teacher*. See For tutors/lecturers in EAP (English for Academic Purposes) BALEAP@JISCMAIL.AC.UK.

Barnett, R. (2015) *Understanding the University: Institution, Idea, Possibilities*. Abingdon: Routledge.

Barnett, R. and Coate, K. (2005) *Engaging the Curriculum in Higher Education*. Maidenhead: Open University Press.

Becher, T. (1994) The significance of disciplinary differences. *Studies in Higher Education* 19 (2), 151–161.

Becher, T. and Trowler, P.R. (2001) *Academic Tribes and Territories: Intellectual Enquiry and the Culture of Disciplines* (2nd edn). Buckingham: The Society for Research into Higher Education and Open University Press.

Beelan, J. and Jones, E. (2015) Redefining internationalization at home. In A. Curaj, L. Matei, R. Pricopie, J. Salmi and P. Scott (eds) *The European Higher Education Area*. See https://www.academia.edu/20566727/Redefining_Internationalisation_at_Home.

Belcher, D.D. (2006) English for specific purposes: Teaching to perceived needs and imagined futures in worlds of work, study and everyday life. *TESOL Quarterly* 40 (1), 133–156.

Benesch, S. (2001) *Critical English for Academic Purposes: Theory, Politics, and Practice*. London: Blackwell Publishing Ltd.

Benzie, H.J. (2010) Graduating as a 'native speaker': International students and English language proficiency in higher education. *Higher Education Research and Development* 29 (4), 447–459.

Berg, M. and Seeber, B.K. (2016) *The Slow Professor: Challenging the Culture of Speed in the Academic*. Toronto: University of Toronto Press.

Bernstein, B. (2000) *Pedagogy, Symbolic Control and Identity: Theory, Research and Critique* (rev. edn). Oxford: Rowman and Littlefield.

Biggs, J. and Tang, C. (2007) *Teaching for Quality Learning at University* (3rd edn). Berkshire: The Society for Research into Higher Education and The Open University Press.

Bleasdale, L. and Humpreys, S. (2017) Undergraduate resilience research project report. Leeds Institute for Teaching Excellence. See https://teachingexcellence.leeds.ac.uk/wp-content/uploads/sites/89/2018/01/LITEbleasdalehumphreys_fullreport_online.pdf.

Boden-Galvez, J. and Ding, A. (2019) Interdisciplinary EAP: Moving beyond aporetic English for general academic purposes. *The Language Scholar.* Issue 4, 78–88.

Bond, B. (2017a) The E(A)P of spelling: Using exploratory practice to (re)engage teachers and students. In J. Kemp (ed.) *Proceedings of the 2015 BALEAP Conference. EAP in a rapidly changing landscape: Issues, challenges and solutions.* Reading: Garnet.

Bond, B. (2017b) Co-constructing the curriculum through exploratory practice. *The Language Scholar* Issue 2, 7–17.

Bond, B. (2018) The Long Read: Inclusive teaching and learning for international students. Leeds: Leeds Institute for Teaching Excellence. See http://teachingexcellence.leeds.ac.uk/the-long-read-inclusive-teaching-and-learning-for-international-students/.

Bond, B. (2019) International students: language, culture and the 'performance of identity'. *Teaching in Higher Education* 24 (5), 649–655.

Bond, B. and Whong, M. (2017) *A combined offer: Collaborative development through a content-based presessional programme.* Presentation at the BALEAP 2017 Conference: Addressing the state of the union: Working together = learning together. Bristol. Available online https://www.baleap.org/event/addressing-state-union-working-together-learning-together.

Borg, S. (2009) English language teachers' conceptions of research. *Applied Linguistics* 30 (3), 358–388.

Borg, S. (2013) *Teacher Research in Language Teaching: A Critical Analysis*. Cambridge: Cambridge University Press.

Bothwell, E. (2017) See https://www.timeshighereducation.com/world-university-rankings/europe-university-rankings-2017-international-hot-spots.

Bourdieu, P. (1991) *Language and Symbolic Power.* Cambridge: Blackwell.

Bourdieu, P. and Wacquant, L. (1992) *An Invitation to Reflexive Sociology*. Cambridge: Polity Press.

Bourdieu, P., Passeron, J.C. and de Saint Martin, M. (1994) *Academic Discourse: Linguistic Misunderstanding and Professorial Power*. Cambridge: Polity Press.

Bovill, C. and Woolmer, C. (2018) How conceptualisations of curriculum in higher education influence student-staff co-creation in and of the curriculum. *Higher Education*. published online 26/12/2018.

Bretag, T. (2017) Assessment design won't stop cheating but our relationships with students might. *The Conversation* (accessed March 2019). https://theconversation.

com/assessment-design-wont-stop-cheating-but-our-relationships-with-students-might-76394.

Brewer, S., Standring, A. and Stansfield, G. (eds) (2019) *Papers from the Professional Issues Meeting (PIM) on Insessional English for Academic Purposes held at London School of Economics 19 March 2016*. Renfrew: BALEAP: The Global Forum for EAP Professionals [accessed online 15/11/2019] https://www.baleap.org/wp-content/uploads/2019/10/Baleap_Book_Interactive.pdf.

Brinton, D.M. and Holten, C.A. (2001) Does the emperor have no clothes? A re-examination of grammar in content-based instruction. In J. Flowerdew and M. Peacock (2001) (eds) *Research Perspectives on English for Academic Purposes* (pp. 239–251). Cambridge: Cambridge University Press.

British Council (2014) *Integration of International Students. A UK Perspective*. British Council: Education Intelligence https://www.britishcouncil.org/sites/default/files/oth-integration-report-september-14.pdf.

British Educational Research Association [BERA] (2018) Ethical Guidelines for Educational Research, fourth edition, London. https://www.bera.ac.uk/researchers-resources/publications/ethicalguidelines-for-educational-research-2018.

Brooke, M. (2019) *Making knowledge more explicit in the English for Academic Purposes classroom*. The Society for Research into Higher Education. See https://srheblog.com/2019/03/22/making-knowledge-more-explicit-in-the-english-for-academic-purposes-classroom/.

Brown, S. (2015) Assessing well at Masters level. In P. Kneale (ed.) *Masters Level Teaching, Learning and Assessments. Issues in Design and Delivery* (pp. 173–182). London: Palgrave Macmillan.

Bruce, I. (2005) Syllabus design for general EAP writing courses: A cognitive approach. *Journal of English for Academic Purposes* 4 (3), 239–256.

Bruce, I. (2008) *Academic Writing and Genre: A Systematic Analysis*. London: Continuum.

Bruce, I. (2011) *Theory and Concepts of English for Academic Purposes*. Basingstoke: Palgrave Macmillan.

Burns, A. (2010) *Doing Action Research in Language Teaching. A Guide for Practitioners*. Abingdon: Routledge.

Cameron, D. (1995) *Verbal Hygiene*. London: Routledge.

Campion, G.C. (2016) 'The learning never ends': Exploring teachers' views on the transition from General English to EAP. *Journal of English for Academic Purposes* 23, 59–70.

Canagarajah, S. (2011) Codemeshing in academic writing: identifying teachable strategies of translanguaging source. *The Modern Language Journal* 95 (3), 401–417.

Canagarajah, S. (2013) *Translingual Practice: Global Englishes and Cosmopolitan Relations*. London: Routledge.

Caplan, N. (2019) *Grammar Choices for Graduate and Professional Writers* (2nd edn). University of Michigan Press.

Carroll, J. (2015) *Tools for Teaching in an Educationally Mobile World*. London: Routledge.

Castells, M. (1996) *The Rise of the Network Society*. Oxford: Blackwell.

Chanock, K. (2007) What academic language and learning advisers bring to the scholarship of teaching and learning: Problems and possibilities for dialogue with the disciplines. *Higher Education Research and Development* 26 (3), 269–280.

Clarence, S. (2016) Exploring the nature of disciplinary teaching and learning using legitimation code theory semantics. *Teaching in Higher Education* 21 (2), 123–37.

Clarke, D.F. (1991) The negotiated syllabus: What is it and how is it likely to work? *Applied Linguistics* 12 (1), 13–28.

Common European Framework of Reference for Languages (2001) Cambridge University Press. See http://www.coe.int/en/web/common-european-framework-reference-languages/.

Cooper, J. (2017) How do we show international students they're still welcome in the UK? *The Guardian*, 17 May 2017.

Cohen, L., Manion, L. and Morrison, K. (2007) *Research Methods in Education* (6th edn). London: Routledge.
Cottingham, M.D. (2016) Theorizing emotional capital. *Theory and Society* 45, 451–470 https://doi.org/10.1007/s11186-016-9278-7.
Davies, A. (2003) *The Native Speaker: Myth and Reality.* Clevedon: Multilingual Matters.
Davis, M. (2019) Publishing research as an EAP practitioner: Opportunities and threats. *Journal of English for Academic Purposes* 39, 72–86.
DEEWR. (2009) *Good practice principles for English language proficiency for international students in Australian universities.* Canberra: Department of Education, Employment and Workplace Relations. Accessed online: http://www.aall.org.au/sites/default/files/Final_Report-Good_Practice_Principles2009.pdf.
Denzin, N.K. and Lincoln, Y.S. (eds) (2005) *The Sage Handbook of Qualitative Research.* (3rd edn). Thousand Oaks: Sage.
Department for Education; UK Government (2017) *Inclusive Teaching and Learning in Higher Education as a route to Excellence.* See https://assets.publishing.service.gov.uk/government/uploads/system/uploads/attachment_data/file/587221/Inclusive_Teaching_and_Learning_in_Higher_Education_as_a_route_to-excellence.pdf.
Ding, A. (2016) Challenging scholarship: A thought piece. *The Language Scholar* 0, 6–19.
Ding, A. (2019a) 'Academic language is … no one's mother tongue': Misusing Bourdieu and a 'morally questionable' Hyland'. *Teaching EAP: Polemical. Questioning, debating and exploring issues in EAP* Blog. See https://teachingeap.wordpress.com/2019/11/01/academic-language-is-no-ones-mother-tongue-misusing-bourdieu-and-a-morally-questionable-hyland/.
Ding, A. (2019b) EAP practitioner identity. In K. Hyland and L.L.C Wong (eds) *Specialised English. New Directions in ESP and EAP Research and Practice* (pp. 69–77). London: Routledge.
Ding, A. and Bruce, I. (2017) *The English for Academic Purposes Practitioner: Operating on the Edge of Academia.* London: Palgrave Macmillan.
Ding, A. and Campion, G. (2016) EAP teacher development. In K. Hyland and P. Shaw (eds) *The Routledge Handbook of English for Academic Purposes* (pp. 571–583). London: Routledge.
Ding, A., Boden-Galvez, J., Bond, B., Morimoto, K., Ragni, V., Rust, N. and Soliman, R. (2018) Manifesto for the scholarship of language teaching and learning. *The Language Scholar* 3, 58–60.
Douglas Fir Group (2016) A transdisciplinary framework for SLA in a multilingual world. *Modern Language Journal* 100(Supplement 2016), 19–47. doi:10.1111/modl.12301.
Dudley-Evans, T. (1997) Five questions for LSP teacher training. In R. Howard and G. Brown (eds) *Teacher Education for LSP* (pp. 58–67). Clevedon: Multilingual Matters.
Duff, P. (2012) Identity, agency and second language acquisition. In S.M. Gass and A. Mackey (eds) *The Routledge Handbook of Second Language Acquisition.* Abingdon: Routledge.
Dyches, J. and Boyd, A. (2017) Foregrounding equity in teacher education: Toward a model of social justice pedagogical and content knowledge. *Journal of Teacher Education* 68 (5), 476–490.
Elbaz, F. (1983) *Teacher Thinking: A Study of Practical Knowledge.* London: Routledge. https://doi.org/10.4324/9780429454615.
Elton, L. (2010) Academic writing and tacit knowledge. *Teaching in Higher Education* 15 (2), 151–160.
Eraut, M. (2004) The practice of reflection. *Learning in Health and Social Care* 3 (2), 47–52. https://doi.org/10.1111/j.1473-6861.2004.00066.x.
Fanghanel, J., McGowan, S., Parker, P., McConnell, C., Potter, J., Locke, W. and Healey, M. (2015) Literature Review. *Defining and Supporting the Scholarship of Teaching and Learning (SoTL): A Sector-Wide Study.* York: Higher Education Academy.

Feak, C. (2016) EAP support for post-graduate students. In K. Hyland and P. Shaw (eds) *The Routledge Handbook of English for Academic Purposes* (pp. 498–501). Abingdon: Routledge.

Felten, P. (2013) Principles of good practice in SoTL. *Teaching and Learning Inquiry* 1 (1), 121–125.

Flowerdew, L. (2015) Adjusting pedagogically to an ELF world: An ESP perspective. In Y. Bayyurt and S. Akcan (eds) *Current Perspectives on Pedagogy for English as a Lingua Franca* (pp. 13–34). Berlin: De Gruyter Mouton.

Flowerdew, J. and Peacock, M. (2001a) (eds) *Research Perspectives on English for Academic Purposes*. Cambridge: Cambridge University Press.

Flowerdew, J. and Peacock, M. (2001b) The EAP curriculum: Issues, methods and challenges. In J. Flowerdew and M. Peacock (eds) *Research Perspectives on English for Academic Purposes*. Cambridge: Cambridge University Press.

Four Corners (2019) Cash cows: Australian universities making billions out of international students. *ABC News*. See https://www.youtube.com/watch?time_continue=4&v=Sm6lWJc8KmE&feature=emb_title.

Freeman, D. (1991) 'To make the tacit explicit': Teacher education, emerging discourse, and conceptions of teaching. *Teaching and Teacher Education* 7, 439–454.

Freire, P. (1996) *Pedagogy of the Oppressed*. (New rev. edn translated by Myra Bergman Ramos). Harmondsworth: Penguin.

Forbes-Mewett, H. and Sawyer, A-M. (2016) International students and mental health. *Journal of International Students* 6 (3), 661–667.

Fulcher, G. (2012) Assessment literacy for the language classroom. *Language Assessment Quarterly* 9 (2), 113–132.

Fung, D. (2017a) *A Connected Curriculum for Higher Education*. London: UCL Press.

Fung, D. (2017b) Strength-based scholarship and good education: The scholarship circle. *Innovations in Education and Teaching International* 54 (2), 101–110.

Gao, F. (2011) Exploring the reconstruction of Chinese learners' national identities in their English-language-learning journeys in Britain. *Journal of Language, Identity and Education* 10 (5), 287–305.

Gardner, S., Nesi, H. and Biber, D. (2018) Discipline, level, genre: Integrating situational perspectives in a new MD analysis of university student writing. *Applied Linguistics* 1–30.

Gee, J.P. (2008) *Social Linguistics and Literacies: Ideology in Discourses* (3rd edn). London: Routledge.

Geertsema, J. (2016) Academic development, SoTL and educational research. *International Journal for Academic Development* 21 (2), 122–134.

Geertz, C. (1973) Thick description: toward an interpretive theory of culture. *The Interpretation of Cultures: Selected Essays* (pp. 3–30). New York: Basic Books.

Gimenez, J. (2009) Beyond the academic essay: Discipline-specific writing in nursing and midwifery. *Journal of English for Academic Purposes* 7 (3), 151–164.

Ginther, A. and Elder, C. (2014) A comparative investigation into understandings and uses of the *TOEFL iBT®* Test, the International English Language Testing Service (Academic) Test, and the Pearson Test of English for Graduate Admissions in the United States and Australia: A case study of two university contexts. *ETS Research Report Series*, 2014(2), 1–39. https://doi.org/10.1002/ets2.12037.

The Guardian [online] (2019a) Outcry after Duke administrator warns Chinese students to speak English. See https://www.theguardian.com/us-news/2019/jan/28/duke-university-chinese-students-speak-english-email-backlash.

The Guardian [online] (2019b) *My University accepts overseas students who are doomed to fail*. Read. See https://www.theguardian.com/education/2019/feb/08/my-university-accepts-overseas-students-who-are-doomed-to-fail?CMP=share_btn_tw.

Glaser, B.G. and Strauss, A.L. (1967) *The Discovery of Grounded Theory: Strategies for Qualitative Research*. New York: Aldine de Gruyte.

Grace, S. and Gravestock, P. (2009) *Inclusion and Diversity. Meeting the Needs of All Students*. Abingdon: Routledge.

Guba, E. and Lincoln, Y. (1994) Competing paradigms in qualitative research. In N.K. Denzin and Y.S. Lincoln (eds) *The Sage Handbook of Qualitative Research* (3rd edn) (pp. 105–117). Thousand Oaks: Sage.

Haber, F. and Griffiths, S. (2017) *5 unique mental health stressors faced by international students*. European Association for International Education blog. See https://www.eaie.org/blog/5-mental-health-stressors-international-students.html.

Hadley, G. (2015) *English for Academic Purposes in Neoliberal Universities: A Critical Grounded Theory*. Heidelberg, Germany: Springer.

Haggis, T. (2006) Pedagogies for diversity: Retaining critical challenge amidst fears of 'dumbing down'. *Studies in Higher Education* 31, 521–535.

Hall, S. (1996) *Modernity: An Introduction to Modern Societies*. Oxford: Blackwell.

Hamp-Lyons, L. (2011) Editorial. *Journal of English for Academic Purposes* 10, 2–4.

Hamp-Lyons, L. and Hyland, K. (2005) Some further thoughts on EAP and JEAP. *Journal of English for Academic Purposes* 4, 1–4.

Hanks, J. (2015) Language teachers making sense of exploratory practice. *Language Teaching Research* 1–22.

Hanks, J. (2017) *Exploratory Practice in Language Teaching: Puzzling About Principles and Practices*. London: Palgrave Macmillan.

Hanks, J. (2019) From research-as-practice to exploratory practice-as-research in language teaching and beyond. *Language Teaching* 52 (2), 143–187. State-of-the-Art article.

Hartig, A.J. (2017) *Connecting Language and Disciplinary Knowledge in English for Specific Purposes: Case Studies in Law*. Bristol: Multilingual Matters.

Harwood, N. (ed.) (2010) *English Language Teaching Materials: Theory and Practice*. Cambridge: Cambridge University Press.

Harwood, N. (2018) What do proofreaders of student writing do to a masters essay? Differing interventions, worrying findings. *Written Communication* 35 (4), 474–530.

Harwood, N. and Hadley, G. (2004) Demystifying institutional practices: Critical pragmatism and the teaching of academic writing. *English for Specific Purposes* 23, 355–377.

HEA (2011) The UK Professional Standards Framework for teaching and supporting learning in higher education. See https://www.heacademy.ac.uk/ukpsf.

HEA (2014) Internationalising Higher Education Framework. See https://www.heacademy.ac.uk/system/files/resources/InternationalisingHEframeworkFinal.pdf.

Healey, M., Flint, A. and Harrington, K. (2014) *Engagement through partnership: students as partners in learning and teaching in higher education* Higher Education Academy. See https://www.heacademy.ac.uk/system/files/resources/engagement_through_partnership.pdf (accessed May 2019).

HM Government (2013) International education – Global growth and prosperity: An accompanying analytical narrative. HM Government. See https://www.gov.uk/government/publications/international-education-strategy-global-growth-and-prosperity (accessed April 2019).

Hockings, C. (2010) *Inclusive Learning and Teaching in Higher education: A Synthesis of Research*. York: Higher Education Academy.

Holliday, A. (1999) Small cultures. *Applied Linguistics* 20 (2), 237–64.

Holliday, A. (2004) Issues of validity in progressive paradigms of qualitative research. *TESOL Quarterly* 38 (4), 731–734.

Holliday, A. (2010) Complexity in cultural identity. *Language and Intercultural Communication* 10 (2), 165–177.

Holliday, A. (2011) *Intercultural Communication and Ideology*. London: Sage.

hooks, b. (1994) *Teaching to Transgress. Education as the Practice of Freedom*. New York: Routledge.

Hutchison, T. and Waters, A. (1987) *English for Specific Purposes: A Learning Centred Approach*. Cambridge: Cambridge University Press.

Hyland, K. (2002) Specificity revisited: How far should we go now? *English for Specific Purposes* 21 (4), 385–395.
Hyland, K. (2004) *Disciplinary Discources: Social Interations in Academic Writing*. Ann Arbor: The University of Michigan Press.
Hyland, K. (2013) Writing in the university: Education, knowledge and reputation. *Language Teaching* 46, 5370.
Hyland, K. (2016b) General and specific EAP. In K. Hyland and P. Shaw (eds) *The Routledge Handbook of English for Academic Purposes* (pp. 17–29). Abingdon: Routledge.
Hyland, K. (2018) First person singular sympathy for the devil? A defence of EAP. *Language Teaching* 51 (3), 383–399.
Hyland, K. and Shaw, P. (eds) (2016) *The Routledge Handbook of English for Academic Purposes*. Abingdon: Routledge.
Hyland, K. and Tse, P. (2007) Is there an 'academic vocabulary'? *TESOL Quarterly* 41 (2), 235– 254.
IELTS (2017) See https://www.ielts.org/ielts-for-organisations/setting-ielts-entry-scores.
Ippolito, K. (2007) Promoting intercultural learning in a multicultural university: Ideals and realities. *Teaching in Higher Education* 12 (5–6), 749–763.
Ivanic, R. (1998) *Writing and Identity: The Discoursal Construction of Identity in Academic Writing*. Philadelphia, PA: John Benjamins.
Jenkins, J. (2013) *English as a Lingua Franca in the International University. The Politics of Academic English Language Policy*. Oxon: Routledge.
Jenkins, J. (2015) Repositioning English and multilingualism in English as a Lingua Franca. *Englishes in Practice* 2 (3), 49–85.
Kachru, B. (1982) *The Other Tongue: English Across Cultures*. Urbana: University of Illinois Press.
Kemmis, S. (2009) Action research as a practice based practice. *Educational Action Research* 17, 463–474.
Kemmis, S. and Smith, T.J. (2008) *Enabling Praxis: Challenges for Education*. Rotterdam: Sense.
Kettle, M. (2017) *International Student Engagement in Higher Education: Transforming Practices, Pedagogies and Participation*. Bristol: Multilingual Matters.
King, N. (2004) *Template analysis – what is template analysis?* See http://www.hud.ac.uk/hhs/research/template-analysis/.
Kirk, S.E. (2018) Enacting the curriculum in English for academic purposes: A legitimation code theory analysis. Ed.D Thesis, Durham University. See http://etheses.dur.ac.uk/12942/.
Kramsch, C. (1993) *Context and Culture in Language Teaching*. Oxford: Oxford University Press.
Kramsch, C. (1998) *Language and Culture*. Oxford: Oxford University Press.
Kramsch, C. (2009) *The Multilingual Subject*. Oxford: Oxford University Press.
Kreber, C. (ed.) (2009) *The University and its Disciplines: Teaching and Learning Within and Beyond Disciplinary Boundaries*. Abingdon: Routledge.
Kumaravadivelu, B. (1994) The postmethod condition: (E)merging strategies for second/foreign language teaching. *TESOL Quarterly* 28 (1), 27–48.
Kumaravadeivelu, B. (2012) *Language eacher ducation for a lobal ociety: A odular odel for nowing, nalyzing, ecognizing, oing and eeing*. New York: Routledge.
Land, R., Rattray, J. and Vivian, P. (2014) Learning in the liminal space: A semiotic approach to threshold concepts. *Higher Education* 67, 199–217.
Lantolf, J.P. and Swain, M. (2019) On the emotion–cognition dialectic: A sociocultural response to prior. *The Modern Language Journal* 103, 528–531.
Lave, J. and Wenger, E. (1991) *Situated Learning: Legitimate Peripheral Participation*. Cambridge: Cambridge University Press.
Law, J. (2004) *After Method. Mess in Social Science Research*. London: Routledge.
Lawrie, G., Marquis, E., Fuller, E., Newman, T., Qui, M., Nomikoudis, M., Roelofs, F. and van Dam, L. (2017) Moving towards inclusive learning and teaching: A synthesis of recent literature. *Teaching and Learning Inquiry* 5 (1).

Lea, M.R. (2004) Academic literacies: A pedagogy for course design. *Studies in Higher Education* 29 (6), 739–756.
Lea, M.R. and Street, B. (1998) Student writing in higher education: An academic literacies approach. *Studies in Higher Education* 23 (2), 157–172.
Leaske, B. (2013) Internationalizing the curriculum in the disciplines – imagining new possibilities. *Journal of Studies in International Education* 17 (2), 103–118.
Leftstan, A. and Snell, J. (2014) *Better than Best Practice: Developing Learning and Teaching through Dialogue*. London, England, Routledge.
Leibowitz, B. and Bozalek, V. (2016) The scholarship of teaching and learning from a social justice perspective. *Teaching in Higher Education* 21 (2), 109–122.
Li, W. (2018) Translanguaging as a practical theory of language. *Applied Linguistics* 39 (1), 9–30.
Lillis, T. (2003) Student writing as 'academic literacies': Drawing on Bakhtin to move from critique to design. *Language and Education* 17 (3), 192–207.
Lillis, T. and Tuck, J. (2016) Academic Literacies: A critical lens on writing and reading in the academy. In K. Hyland and P. Shaw (eds) *The Routledge Handbook of English for Academic Purposes* (pp. 30–43). Abingdon: Routledge.
Lillis, T. and Turner, J. (2001) Student writing in higher education: Contemporary confusion, traditional concerns. *Teaching in Higher Education* 6 (1), 57–68.
Lillis, T., Harrington, K., Lea M.R. and Mitchell, S. (2015) *Working with Academic Literacies: Case Studies Towards Transformative Practice*. The WAC Clearinghouse, Fort Collins, Colorado 80523-1052.
MacDonald, J. (2016) The margins as third space: EAP teacher professionalism in Canadian Universities. *TESL Canada Journal* 31 (1).
MacLean, M., and Poole, G. (2010) An introduction to ethical considerations for novices to research in teaching and learning in Canada. *The Canadian Journal for the Scholarship of Teaching and Learning* 1 (2).
Margolis, R. (2016) Exploring internationalisation and the international student identity. *The Language Scholar* 0, 49–67.
Martin, R.C. (2013) Navigating the IRB: The ethics of SoTL. *New Directions for Teaching and Learning* 136, 59–71.
Maton, K. (2014) *Knowledge and Knowers: Towards a Realist Sociology of Education*. London: Routledge.
Maton, K. and Muller, (2007) A sociology for the transmission of knowledges. In F. Christie and J. Martin (eds) *Language, Knowledge and Pedagogy: Functional Linguistic and Sociological Perspectives* (pp. 14–33). London: Continuum.
Matsuda, P.K. (2014) The lure of translingual writing. *PMLA* 129 (3), 478–483. https://doi.org/10.1632/pmla.2014.129.3.478.
Mauranen, A. (2003) The corpus of English as a Lingua Franca in academic settings. *TESOL Quarterly* 37 (3), 513–27.
Mauranen, A. (2012) *Exploring ELF. Academic English Shaped by Non-native Speakers*. Cambridge: Cambridge University Press.
Mauranen, A., Hynninen, N. and Ranta, E. (2016) English as the academic lingua franca. In K. Hyland and P. Shaw (eds) *The Routledge Handbook of English for Academic Purposes* (pp. 44–55). Abingdon: Routledge.
May, H. and Bridger, K. (2010) *Developing and Embedding Inclusive Policy and Practice in Higher Education*. York: The Higher Education Academy.
McIntosh, K., Connor, U. and Gokpinar-Shelton, E. (2017) What intercultural rhetoric can bring to EAP/ESP writing studies in an English as a lingua franca world. *Journal of English for Academic Purposes* 29, 12–20. http://dx.doi.org/10.1016/j.jeap.2017.09.001.
McWilliam, E. (2005) Unlearning pedagogy. *Journal of Learning Design* 1 (1), 1–11.
McWilliam, E. (2008) Unlearning how to teach. *Innovations in Education and Teaching International* 45 (3), 263–269.

Meddings, L. and Thornbury, S. (2009) *Teaching Unplugged: Dogme in English Language Teaching*. Peaslake: Delta Publishing.

Meyer, J. and Land, R. (2003) Threshold concepts and troublesome knowledge: Linkages to ways of thinking and practising within the disciplines. *Enhancing Teaching-Learning Environments in Undergraduate Courses Occasional Report* 4, ETL Project, Universities of Edinburgh, Coventry and Durham.

Meyer, J. and Land, R. (eds) (2006) *Overcoming Barriers to Student Understanding: Threshold Concepts and Troublesome Knowledge*. Abingdon: Routledge.

Meyer, J.H.F. and Land, R. (2005) Threshold concepts and troublesome knowledge (2): Epistemological considerations and a conceptual framework for teaching and learning. *Higher Education* 49 (3). Issues in Teaching and Learning from a Student Learning Perspective: A Tribute to Noel Entwhistle 373–388 Springer.

McKie, A. (2019) International students and cheating: how worried should we be? *The Times Higher Education*. See https://www.timeshighereducation.com/news/international-students-and-cheating-how-worried-should-we-be.

Monbec, L. (2018) Designing an EAP curriculum for transfer: A focus on knowledge. *Journal of Academic Language and Learning* 12 (2), 88–101.

Montgomery, C. (2010) *Understanding the International Student Experience*. Basingstoke: Palgrave Macmillan.

Moore, R. (2007) Going critical: The problem of problematizing knowledge in education studies. *Critical Studies in Education* 48 (1), 25–41.

Mora, J.M. (2017) Universities must value their students more than their reputation. *The Times Higher Education*. See https://www.timeshighereducation.com/blog/universities-must-value-their-students-more-their-reputation.

Moran, J. (2017) See https://www.theguardian.com/higher-education-network/2017/jun/07/top-200-universities-in-the-world-2016-the-uks-rise-and-fall 7 June 2017.

Moran, M. (2018) (Un)troubling identity politics: A cultural materialist intervention. *European Journal of Social Theory* 1–20.

Morgan, B. (2009) Fostering transformative practitioners for critical EAP: Possibilities and challenges. *Journal of English for Academic Purposes* 8, 86–99.

Munby, J. (1978) Communicative syllabus design. A sociolinguistic model for defining the content of purpose-specific language programmes. Cambridge: Cambridge University Press.

Mundt, K. and Groves, M. (2015) Friend or foe? Google Translate in language for academic purposes. *English for Specific Purposes* 37, 112–121.

Mundt, K. and Groves, M. (2016) A double-edged sword: the merits and the policy implications of Google Translate in higher education. *European Journal of Higher Education* 6 (4), 387–401.

Murray, N. (2016a) *Standards of English in Higher Education: Issues, Challenges and Strategies*. Cambridge: Cambridge University Press.

Murray, N. (2016b) An academic literacies argument for decentralizing EAP provision. *ELT Journal* 70 (4), 435–443.

Nesi, H. and Gardner, S. (2012) *Genre Across the Disciplines: Student Writing in Higher Education*. Cambridge University Press.

Nesi, H. and Hu, G. via BALEAP JISCmail (2019) *'Saturday Events'*. See For tutors/lecturers in EAP (English for Academic Purposes) BALEAP@JISCMAIL.AC.UK.

Norton, B. (2000) *Identity and Language Learning: Gender, Ethnicity and Educational Change*. Harlow: Pearson Education.

Norton, B. (2016) Identity and language learning: Back to the future. *TESOL Quarterly* 50 (2), 475–479.

Norton Peirce, B. (1995) Social identity, investment, and language learning. *TESOL Quarterly* 29 (1), 9–31.

Ortactepe, D. (2013) 'This Is called free-falling theory not culture shock!': A narrative inquiry on second language socialization. *Journal of Language, Identity and Education* 12 (4), 215–229.

Pajak, C. (2018) Language for law: A beginner's mind. *The Language Scholar* 3, 33–42.
Paxton, M. and Frith, V. (2014) Implications of academic literacies research for knowledge making and curriculum design. *Higher Education* 67, 171–182. DOI 10.1007/s10734-013-9675-z.
Pennycook, A. (1996) Borrowing others' words: Text, ownership, memory and plagiarism. *TESOL Quarterly* 30 (2), 201–230.
Pennycook, A. (1997) Vulgar pragmatism, critical pragmatism and EAP. *English for Specific Purposes* 16 (4), 253–269.
Perkins, D. (1999) The many faces of constructivism. *Educational Leadership* 57 (3), 6–11.
Pilcher, N. and Richards, K. (2017) Challenging the power invested in the International English Language Testing System (IELTS): Why determining 'English' preparedness needs to be undertaken within the subject context. *Power and Education* 9 (1), 3–7.
Porpora, D.V. (2015) *Reconstructing Sociology: The Critical Realist Approach*. Cambridge: Cambridge University Press.
Potter, M.K. and Kustra, E.D.H. (2011) The relationship between scholarly teaching and SoTL: Models, distinctions, and clarifications. *International Journal for the Scholarship of Teaching and Learning* 5 (1), Article 23.
Race, P. (2015) Assessment digest: Exams, essays and more? See https://phil-race.co.uk/assessment/ (accessed 8 May 2019).
Raimes, A. (1991) Instructional balance: From theories to practices in the teaching of writing. In J. Alatis (ed.) *Georgetown University Round Table on Language and Linguistics*. Washington, DC: Georgetown University Press.
Richards, J.C. (2013) Curriculum approaches in language teaching: Forward, central, and backward design. *RELC Journal* 44 (1), 5–33.
Richards J.C. and Rodgers T. (2001) *Approaches and Methods in Language Teaching* (2nd edn). New York: Cambridge University Press.
Roxå, T. and Mårtensson, K. (2009) Significant conversations and significant networks – exploring the backstage of the teaching arena. *Studies in Higher Education* 34 (5), 547–559.
Roxå, T. and Mårtensson, K. (2016) Peer engagement for teaching and learning: Competence, autonomy and social solidarity in academic microcultures. *Uniped* 39 (2), 131–143 www.idunn.no/uniped.
Ryan, J. (2011) Teaching and learning for international students: Towards a transcultural approach. *Teachers and Teaching* 17 (6), 631–648.
Schon, D. (1983) *The Reflective Practitioner: How Professionals Think in Action*. New York: Basic Books.
Seburn, T. (2011) *Academic Reading Circles*. CreateSpace Independent Publishing Platform.
Seidlhofer, B. (2011) *Understanding English as a Lingua Franca*. Oxford: Oxford University Press.
Shulman, L.S. (1987) Knowledge and teaching: Foundations of the new reform. *Harvard Educational Review* 1–22.
Shulman, L.S. (2000) From Minsk to Pinsk: Why a scholarship of teaching and learning? *The Journal of Scholarship of Teaching and Learning* 1, 48–52.
Silverman, D. (2000) *Doing Qualitative Research* (1st edn). London: Sage.
Simpson, M. and Tuson, J. (2003) *Using Observations in Small-Scale Research*. Glasgow: SCRE, University of Glasgow.
Sloan, D. and Porter, E. (2010) Changing international students' and business staff perceptions of in-sessional EAP: Using the CEM model. *Journal of English for Academic Purposes* 9 (3), 198–210.
Smith, R. and Rebolledo, P. (2018) *A Handbook for Exploratory Action Research*. London: British Council. Open Access.
Spencer-Oatey, H. (2019) Increasing inclusivity in our universities: Why is it important and how are we doing? Inside Government. See http://www.insidegovernment.co.uk/blog/inclusivity-in-universities/.

Spolsky, B. (2012) What is language policy? In B. Spolsky (ed.) *The Cambridge Handbook of Language Policy* (pp. 3–15). Cambridge: Cambridge University Press.

Sterzuk, A. (2015) 'The standard remains the same': language standardisation, race and othering in higher education. *Journal of Multilingual and Multicultural Development* 36 (1), 53–66.

Street, B.V. (1995) *Social Literacies. Critical Approaches to Literacy in Development, Ethnography and Education.* Harlow: Pearson Education.

Swales, J.M. (1990) *Genre Analysis: English in Academic and Research Settings.* Cambridge: Cambridge University Press.

Swales, J.M. (2019) The futures of EAP genre studies: A personal viewpoint. *The Journal of English for Academic Purposes* 38, 75–82.

Tardy, C.M. (2017) Crossing, or creating, divides? A plea for transdisciplinary scholarship. In B. Horner and L. Tetrault (eds) *Crossing Divides: Exploring Translingual Writing Pedagogies and Programs* (pp. 181–189). Logan, UT: Utah State University Press.

Taylor, K. (2018) Temporalities of learning: lessons from a socio-material study of allotment gardening practice. *Studies in Continuing Education* 40 (3), 290–305.

Thomas, L. and May, H. (2010) Inclusive Learning and Teaching in Higher Education. The Higher Education Academy. See https://www.heacademy.ac.uk/system/files/inclusivelearningandteaching_finalreport.pdf.

Thompson, P. (2019) 'Starting the PhD – learning new vocabulary. Patter Blog. See https://patthomson.net/2019/01/28/starting-the-phd-learning-new-vocabulary/ (accessed 11 February 2019).

The Times Higher Education [online], 19 March 2019. See https://www.timeshighereducation.com/news/international-students-and-cheating-how-worried-should-we-be.

Tribble, C. (2017). ELFA vs. genre: A new paradigm war in EAP writing instruction? *Journal of English for Academic Purposes* 25, 30–44.

Tribble, C. (2009) Writing academic English – A survey review. *ELT Journal* 63 (4), 400–417.

Tsui, A.B.M (2003) *Understanding Expertise in Teaching: Case Studies of Second Language Teachers.* Cambridge: Cambridge University Press.

Turner, J. (2004) Language as academic purpose. *Journal of English for Academic Purposes* 3, 95–109.

Turner, J. (2011) *Language in the Academy: Cultural Reflexivity and Intercultural Dynamics.* Bristol: Multilingual Matters.

Turner, J. (2012) Academic literacies: Providing a space for the socio-political dynamics of EAP. *Journal of English for Academic Purposes* 11 (1), 17–25.

Turner, J. (2018) *On Writteness. The Cultural Politics of Academic Writing.* London: Bloomsbury.

UKCISA (UK Council for International Student Affairs) Investigating the mental health and well-being of global access students. Blog Post. See https://ukcisa.org.uk/blog/6597/Investigating-the-mental-health-and-wellbeing-of-global-access-students (accessed 15 February 2019).

UK Government. (2010) Equality act. See http://www.legislation.gov.uk/ukpga/2010/15/contents.

University of Leeds (2017) Code of practice on assessment. See http://ses.leeds.ac.uk/downloads/download/997/code_of_practice_on_assessment_201718.

Universities UK. (2017) Widening participation in UK outward student mobility. See http://www.universitiesuk.ac.uk/policy-and-analysis/reports/Pages/widening-participation-in-uk-outward-student-mobility-a-picture-of-participation.aspx (accessed 2 February 2019).

van Lier, L. (1996) *Interaction in the Language Classroom: Awareness, Autonomy and Authenticity.* Harlow: Longman.

Voloshinov V.N., Matejka, L. and Titunik I.R. (1973) *Marxism and the Philosophy of Language.* New York: Seminar Press.

Watson Todd, R. (2003) EAP or TEAP? *Journal of English for Academic Purposes* 2 (2), 147–156.

Webster-Wright, A. (2009) Reframing professional development through understanding authentic professional learning. *Review of Educational Research* 79 (2), 702–739.
White, C.J. (2018) Agency and emotion in narrative accounts of emergent conflict in an L2 classroom. *Applied Linguistics* 39 (4), 579–598.
Wingate, U. (2015) *Academic Literacy and Student Diversity: The Case for Inclusive Practice*. Bristol: Multilingual Matters.
Wingate, U. (2018) Academic literacy across the curriculum: Towards a collaborative instructional approach. *Language Teaching* 51 (3), 349–364.
Wingate, U. and Tribble, C. (2012) The best of both worlds? Towards an English for academic purposes/academic literacies writing pedagogy. *Studies in Higher Education* 37 (4), 481–495.
Wright, C. (2017) *Creative vs Performative Competence – other orientation in using strategies for second language Academic Interactional Competence*. Centre for Language Education Research Seminar, The University of Leeds 25 January 2017.
Yin, R. (1994) *Case Study Research*. London: SAGE Publications Ltd.
Young, M.F.D. (1998) *The Curriculum of the Future: from the 'New Sociology of Education' to a Critical Theory of Learning*. London: Falmer Press.
Zamel, V. (1998) Strangers in academia: The experience of faculty and ESL students across the curriculum. In V. Zamel and R. Spack (eds) *Negotiating Academic Literacies. Teaching and Learning Across Languages and Cultures* (pp. 249–264). Mahwah, NJ: Lawrence Erlbaum Associates.
Zembylas, M. (2007) Emotional capital and education: Theoretical insights from Bourdieu. *British Journal of Educational Studies* 55 (4), 443–463.
Zepke, N. (2015) Student engagement research: Thinking beyond the mainstream. *Higher Education Research and Development* 34 (6), 1311–1323.

Index

academic integrity 3, 107, 149, 195
plagiarism 3, 48, 113, 120, 172
academic literacy 91, 126, 127
Academic Literacies
and EAP 7, 91, 145, 169, 194, 195
as a framework for inclusive learning and teaching 30, 124, 161, 201
as a research framework x, 21, 26, 68
Advance HE 14, 185
agency 13, 21, 26, 52, 63, 76, 150, 157, 175, 197, 200
abdication of 157–162, 174
and power 138
definition 83–84
in the curriculum 84–87, 88
applied linguistics x, 7, 19, 32, 62, 63, 71
assessment 2, 23, 24, 30, 40, 46, 67, 87, 107
and inclusion 108
and syllabus design 53–55
and time 88, 116
clarity of 58, 59, 70, 81, 84–86, 115, 117, 119, 126, 199
criteria 50, 109–111, 116
genre 115, 119
impact of language in 74, 93, 95, 106–107
literacy 108
of language 38, 39, 53, 92, 109
of writing 181
pre-sessional assessment 8, 36, 39, 159–161, 170

BALEAP 26, 32, 151, 157, 168, 186, 195

CEFR (Common European Framework) 6, 36, 108
CEM Model (Contextual, Embedded, Mapped) 90, 194
codemeshing 48, 184
codeswitching 6
collaboration between EAP and content teachers 28, 29, 33, 46, 91, 93, 125, 145, 157, 180–184, 190, 192, 194–195, 198–199
community of practice 9, 12, 172, 197
confidence
building and EAP 54, 153, 154, 157, 160, 161, 162–166, 173, 180, 191
linguistic 6, 102, 103, 105
students (lack or loss of) 43, 45, 49, 50, 69, 80, 81, 82, 85, 119, 123, 129, 130
teachers (lack or loss of) 39, 75, 77, 79, 169, 170, 171, 192
'corporate agent' 83
corpus linguistics 7, 53, 93, 105, 169
Critical (Social) Realism 18, 21, 53, 62
Critical EAP x, 7, 161, 198
critical thinking 37, 119, 182
cultural norms 20, 138, 148
culture
and language 133, 140, 148
'big' 132–133
educational 113, 116, 127, 129, 134, 135, 147, 183, 197, 198, 200–201
theories of 132

deficit (language) 6, 9, 36, 60, 66, 68, 105, 110, 122–126, 131, 135, 140, 161, 193, 196, 200–201
disciplinary discourse 143, 144–148, 153, 180, 183, 187, 193
discourse analysis x, 101, 169
discourse competence 129, 130
diversity 3, 136, 184–185, 201

English as a Lingua Franca 5, 184, 198, 201
Error correction (language) 39, 44, 50, 94, 109, 111, 112, 139, 182
ethics 11, 15–16, 181, 184
ethnography (ethnographic) 11, 18, 21–22, 26, 174–175

feedback, formative assessment 36, 48, 69, 74, 84, 87, 93, 114, 115, 118–119, 149, 189
fee-paying structures 3, 4, 5, 43, 199

genre 90, 92, 153, 172, 180, 191, 194
genre analysis x, 54
genre theory 7, 170
grammar 37, 56, 93–97, 104, 105, 106, 111, 127, 145, 146, 165, 182, 190

habitus 9, 128, 144, 145, 147, 148, 162, 187
'home' students 23, 58, 59, 89, 95, 99, 124, 131, 137, 141, 200

identity
    and emotion 76
    and Threshold Concepts 143, 147, 148
    cultural 46, 104
    definition 62–63
    disciplinary 29, 71, 76, 89, 144, 158
    of EAP 151–157, 163, 168, 173, 174, 191, 203
    of students 47, 68–70, 88
    of teachers ix, 10, 18, 32, 63–67, 74, 76, 187, 197
    social 137, 138
    sociolinguistics 6, 44, 147
    theories of x, 47, 52, 83–84, 132
IELTS (International English Language Testing System) 6, 809, 35–39, 46, 54, 100, 108, 145–146, 167, 178, 182–183, 201
inclusion 1–3, 51, 89, 107, 125, 148, 179, 188, 195, 201
insessional 27, 32, 37, 41, 46, 73, 90, 153, 157, 194–196
institutional structures 7, 8, 26, 46, 50, 53, 86, 75, 199, 200–201
intercultural communication 130, 136, 152, 157, 176, 179, 189, 191, 194
internationalisation ix, xi, 1–5, 63, 86, 150, 176, 200

Journal of English for Academic Purposes (JEAP) 151

knowledge structures 120, 143

language centres 7, 27–28
language proficiency 36, 38, 42, 79, 95, 100, 105, 108–109, 119, 122, 126, 133, 140, 167, 181, 183, 195–196, 201
language structure 201, 94, 111, 112
Legitimation Code Theory (LCT) 169

liminality (in learning) 21, 143, 147, 148, 172, 179

mechanics of language 39, 94, 95
microcultures 75–79, 187, 197
morphogenesis 53, 84, 137, 177
    elaboration xii, 22, 53, 62, 68, 83, 143

neoliberalism 4, 63, 79, 83, 86, 151, 158

'othering' 82, 88, 130

policy 194–201
    admissions 35
    assessment 112
    education 19, 71
    internationalisation 1, 4, 86, 158
    language 194, 201
power structures x, 6, 137
practitioner research 13, 15, 17, 20
    Action Research 13, 18
    Exploratory Practice 13, 20, 22, 197
pre-sessional 809, 25, 27–28, 32, 36–38, 46, 94–95, 145–146, 153, 158, 161, 165–175, 180, 182, 190
professional learning 13, 17, 77, 168–169, 185–187, 192, 197–198

Quality Assurance 87
    Quality Assurance Agency 56, 61

resistance to cultural/linguistic change 26, 33, 43–46, 62, 64, 66, 104, 138, 142, 143, 147, 148, 152, 171, 181, 186, 189
resources, as capital 8, 28, 48, 58, 70, 137, 141, 147, 148, 155, 167, 186, 196
    linguistic 184, 133
    teaching 91, 127
Russell Group university 27, 29, 61, 64, 74

social justice 2, 5, 8, 108, 200–201
sociolinguistics 6, 108
structure in Social Realism 3, 6, 21, 50, 53, 71, 72, 83–84, 86, 128, 137
support structures 150
Systemic Functional Linguistics 7, 169

teaching excellence 1, 11–12, 14, 18, 27, 33, 132, 202
TEAP Competency Framework, *see* BALEAP

Threshold Concept x, 21, 142–145, 185
Time/temporality 21, 166, 168
 definition of 71–72
 in the curriculum 72–75
 lack of 74, 138, 147, 166
translanguaging 6, 48, 184
translation 41, 44, 48, 113, 146, 149, 184, 196

UKPSF (UK Professional Standards Framework) *see* Advance HE

visa regulations 3, 5, 158
vocabulary 24, 41, 42, 43, 44–50, 92–95, 97–106, 146, 172, 182

widening participation 2, 5, 179

Lightning Source UK Ltd.
Milton Keynes UK
UKHW022014280720
367315UK00003B/75